Managing Production Large Language Models

Designing, Deploying, and Operating LLM Systems at Scale

Jordan O'Neal

Cybersoft Publishing LLC

Fort Washington, MD 20744

Drclaude.net

First Edition, April 2026

1 Contents

Introduction9

1 Introduction and the LLM Production Landscape3

1.1 Opening Scenario3
1.2 Why It Matters5
1.3 What Counts as an LLM: A Taxonomy of Models6
1.4 On-Premises, Cloud, Hybrid, Edge Deployment Models9
1.5 Business Use Cases and an ROI Framework12
1.6 Risks, Failure Modes, and the New Operational Surface15
1.7 Reg. Context: EU AI Act, Sector Rules, Internal Policy18
1.8 Success Metrics and KPIs for Production LLM Projects20
1.9 The Build, Buy, and Compose Decision22
1.10 LLM Project Roles: Architect, Engineer, Lead, and Platform25
1.11 How to Read This Book Companion Repository Overview27
1.12 A Working Definition of 'Production' for LLMs29
1.13 Technical Checklist32
1.14 Team Conversation34
1.15 Key Takeaway35

2 Model Selection and Sizing37

2.1 Opening Scenario37
2.2 Why It Matters38
2.3 Model Families and Their Trade-Offs40
2.4 Open-Source Models versus Managed API Services42
2.5 The Performance, Cost, and Latency Decision Framework45
2.6 Capability Benchmarking Methodology47
2.7 Latency and Throughput Benchmarking49
2.8 Cost Modeling Across Selection Candidates52
2.9 Evaluation Metrics for Production Readiness54
2.10 Multimodal and Reasoning Model Selection56
2.11 Vendor Lock-In, Upgrade Paths, Multi-Model Strategies58
2.12 Sizing Checklist for Infrastructure and Budget Planning61
2.13 Technical Checklist63
2.14 Team Conversation65
2.15 Key Takeaway66

3 Fine-Tuning and Adaptation69

3.1 Opening Scenario69
3.2 Why It Matters71
3.3 Full Fine-Tuning versus Parameter-Efficient Methods72
3.4 LoRA, Adapters, Prefix Tuning, and Modern PEFT Families74
3.5 Cont Pretraining vs Instr. Tuning vs Preference Tuning77
3.6 Data Curation, Labeling, and Augmentation Strategies79
3.7 Building Reproducible Fine-Tuning Pipelines82
3.8 Evaluation and Validation for Adapted Models84
3.9 Synthetic Data and Distillation Patterns86
3.10 Catastrophic Forgetting and Mitigation Patterns88
3.11 Multi-Adapter Serving and Per-Tenant Customization90
3.12 When to Fine-Tune vs When to Prompt vs When to Retr.92
3.13 A Rollback and Deprecation Plan for Adapted Models94
3.14 Technical Checklist96
3.15 Team Conversation98
3.16 Key Takeaway100

4 Prompt Engineering and Prompt Ops102

4.1 Opening Scenario102
4.2 Why It Matters103
4.3 Prompt Design Patterns and Templates105
4.4 Prompt Chaining and Composition107
4.5 Structured Outputs, JSON Mode, and Function Calling109
4.6 Prompt Testing and Versioning Strategies111
4.7 Eval-Driven Development for Prompt Changes114
4.8 Prompt Governance and Review Workflows116
4.9 Cost-Aware Prompting Strategies118
4.10 Prompt and Response Caching120
4.11 Multi-Modal and Multi-Lingual Prompt Patterns122
4.12 Prompt Lifecycle Management124
4.13 Prompt Ops Maturity Model126
4.14 Technical Checklist129
4.15 Team Conversation130
4.16 Key Takeaway131

5 Serving Architectures and Latency Engineering133

5.1 Opening Scenario133
5.2 Why It Matters134
5.3 Real-Time, Batch, and Streaming Serving Patterns136

5.4 Hybrid and Multi-Tier Serving Architectures138
5.5 Model Sharding, Tensor and Pipeline Parallelism for Inference140
5.6 Quantization, Distillation, and Compression Patterns143
5.7 KV Cache Management and Paged Attention145
5.8 Cont Batching and Speculative Decoding in Serving147
5.9 Autoscaling Strategies for LLM Workloads150
5.10 Caching Layers: Result, Semantic, Prefix, Embedding152
5.11 Latency Budgets, TTFT, and Streaming UX Patterns154
5.12 Edge, Browser, and Mobile Serving Patterns156
5.13 Safe-Deploy Pats: Canary, Shadow, Blue-Green, Rollback158
5.14 Technical Checklist160
5.15 Team Conversation162
5.16 Key Takeaway163

6 Cost Engineering and Resource Management166

6.1 Opening Scenario166
6.2 Why It Matters167
6.3 Cost Models for Inference, Fine-Tuning, and Training169
6.4 Unit Economics: Cost Per Request, Per User, Per Feature171
6.5 Spot Instance Strategies and Fallback Policies172
6.6 Request Batching and Shaping Techniques174
6.7 Model-Tiering and Cascade Patterns176
6.8 Caching as a Cost Strategy178
6.9 Monitor CPR, Real-Time Cost Anomaly Detection180
6.10 Chargeback, Showback Models for Multi-Team Envs182
6.11 Reservation, Commitment, and Capacity Plan. Strategies184
6.12 FOSS, Self-Hosted, and Hybrid Cost Profiles185
6.13 A Cost Control Playbook with Tactics in Priority Order187
6.14 Technical Checklist190
6.15 Team Conversation191
6.16 Key Takeaway193

7 Safety, Security, and Compliance195

7.1 Opening Scenario195
7.2 Why It Matters197
7.3 Adversarial Inputs and Prompt Injection Defense198
7.4 Hallucination Detection and Mitigation200
7.5 Output Filtering, Guardrails, and Policy Enforcement202

7.6 Access Control and Authentication for LLM APIs204
7.7 Data Privacy, PII Handling, and Memorization Risk206
7.8 Logging Policies and Audit Trails for LLM Interactions208
7.9 Adversarial Red-Teaming and Continuous Testing210
7.10 Watermarking, Provenance, and Content Authenticity212
7.11 Regulatory Compliance Frameworks214
7.12 Incident Response for Model Failures216
7.13 A Safety Checklist for Production LLM Systems218
7.14 Technical Checklist219
7.15 Team Conversation221
7.16 Key Takeaway222

8 Observability and Monitoring for LLMs224

8.1 Opening Scenario224
8.2 Why It Matters225
8.3 Key Metrics: Latency, TTFT, Token Usage, and Cost226
8.4 Quality Metrics: Faithfulness, Helpfulness, Refusal Rates229
8.5 Semantic Drift and Behavioral Drift Detection230
8.6 Logging Strategies for LLM Requests and Responses232
8.7 Distributed Tracing for Multi-Step Pipelines234
8.8 User Feedback Loops and Quality Signals236
8.9 Real-Time Evaluators and LLM-as-Judge in Production237
8.10 Alerting, SLOs, and SLA Design for LLM Systems239
8.11 Cost Observability and Anomaly Detection241
8.12 Postmortems Root-Cause Analysis for Model Incidents243
8.13 An Observability Maturity Model for LLM Platforms245
8.14 Technical Checklist247
8.15 Team Conversation249
8.16 Key Takeaway250

9 Testing, CI/CD, and Model Validation252

9.1 Opening Scenario252
9.2 Why It Matters253
9.3 Unit Tests for Prompts and Model Outputs255
9.4 Eval Suites: Golden Sets, Rubrics, and LLM-as-Judge257
9.5 Integration Testing Strategies for LLM Pipelines259
9.6 Traffic-Replay Testing and Production-Shape Evaluation261
9.7 Canary Releases and Blue-Green Deployments263
9.8 Shadow Deployments for Quality Comparison265

9.9 Reproducible Training and Adaptation Pipelines267
9.10 Artifact Storage and Model Registry Patterns269
9.11 Continuous Evaluation in Production271
9.12 Release Engineering for LLM Applications273
9.13 A CI/CD Checklist for LLM Applications275
9.14 Technical Checklist278
9.15 Team Conversation279
9.16 Key Takeaway280

10 Human-in-the-Loop and Feedback Systems282

10.1 Opening Scenario282
10.2 Why It Matters283
10.3 Labeling Pipelines and Active Learning284
10.4 RLHF, DPO, and Preference-Tuning Patterns287
10.5 Human Review Workflows and Escalation Paths288
10.6 UX Patterns for Feedback Collection290
10.7 Implicit Feedback Signals: Regen, Copy, Edit, Follow-Up292
10.8 Feedback Calibration and Rater Quality Control294
10.9 Closing the Loop: From User Signal to Model Update296
10.10 Privacy-Preserving Feedback Collection298
10.11 Cost and Latency Trade-Offs of Human Review299
10.12 Retiring Human Review: When the Model Is Ready301
10.13 Technical Checklist304
10.14 Team Conversation306
10.15 Key Takeaway307

11 RAG, Agents, and Tool-Using LLM Systems in Prod309

11.1 Opening Scenario309
11.2 Why It Matters310
11.3 RAG, Tool-Calling, Agentic: 3 Shapes Modrn LLM Apps312
11.4 Retrieval Qual: Recall, Precision, and Relevance Patterns315
11.5 Embedding Models and Vector Index Operations317
11.6 Chunking, Document Parsing, and Knowledge Curation319
11.7 Hybrid Search, Re-Ranking, and Fusion Patterns321
11.8 Tool Calling and Function-Calling Patterns324
11.9 Agent Loop Patterns: ReAct, Plan-Execute, Hierarchical326
11.10 Multi-Agent Systems and Orchestration Patterns328
11.11 Cost Ceilings, Step Budgets, Termination Conditions330
11.12 Memory, State, and Long-Running Agent Patterns332

11.13 Observability for RAG and Agentic Systems334
11.14 Safety, Sandboxing, Permission for Tool-Using Agents336
11.15 Technical Checklist348
11.16 Team Conversation350
11.17 Key Takeaway351

12 Case Studies, Industry Patterns, Reference Arch353

12.1 Opening Scenario353
12.2 Why It Matters354
12.3 Case Study: Financial Services LLM Deployment356
12.4 Case Study: Healthcare NLP Pipeline359
12.5 Case Study: E-Commerce Recommendation and Search369
12.6 Case Study: Enterprise Productivity Assistant372
12.7 Cross-Case Patterns and Lessons Learned376
12.8 An Anti-Pattern Catalog for Production LLM Systems377
12.9 Industry-Specific Compliance Considerations380
12.10 Ref-Arch Template for Production LLM Sys383
12.11 A Maturity Model for Production LLM Capability385
12.12 A Ninety-Day Plan to Advance One Maturity Stage388
12.13 Reference Architecture: Multi-Tenant LLM Platform390
12.14 Technical Checklist393
12.15 Team Conversation394
12.16 Key Takeaway395

2 Introduction

The RAG-powered contract assistant had been live for eleven days when the first complaint arrived. A procurement officer at a mid-size enterprise had asked the assistant to confirm the penalty clause in a vendor agreement. The assistant responded with confident, fluent prose, cited the correct vendor name, and stated the penalty cap at 2% of the annual contract value. The actual clause read five percent. The officer, trusting the system, forwarded the answer to the vendor's account team and used it as the basis for a negotiation position. The error surfaced three days later during legal review.

The engineering post-mortem traced the failure to four overlapping causes. The retrieval layer had returned a document from a prior contract version because the vector index had not been refreshed after the amendment was signed. The prompt did not instruct the model to cite the specific clause number. The model, trained on a broad corpus of contract language, had a strong prior toward the two-percent figure and confidently filled the gap. The evaluation suite the team ran before launch tested general question-answering accuracy but did not include adversarial document-version tests. No monitoring alert was in place to flag responses that contradicted the source document. The team had built the prototype in two sprints, demoed it to enthusiastic applause, and shipped it to thirty users without closing any of these gaps. The system behaved exactly as it was designed: it generated fluent, plausible text grounded in whatever the retriever returned. The retriever returned the wrong document.

The same week, a separate team at the same company discovered that an autonomous email-drafting agent had entered a retry loop after an upstream API returned a 429 rate-limit response. The agent's orchestration layer caught the error, waited, and retried. The upstream API continued to throttle. The agent retried 43 times over a 90-minute window, generating a new draft email for each attempt, logging it, and attempting delivery. No human was in the loop. No budget cap had been set on the agent's token spend. By the time the on-call engineer saw the alert, the agent had consumed four times its daily token budget and had written forty-three near-identical drafts to the same recipient's outbox queue. The recipient was a regulator.

These two incidents, assembled from patterns that recur across organizations and deployment shapes, name the central difficulty of putting large language models into production: LLM systems do not fail the way classical software fails. They fail quietly, confidently, and in ways that look correct until someone checks the source. The retrieval system returned a document. The model generated text. Every component did what it was designed to do. The failure was not in the code; it was in the absence of the operational disciplines that production-grade LLM systems require. This book is the manual for those disciplines.

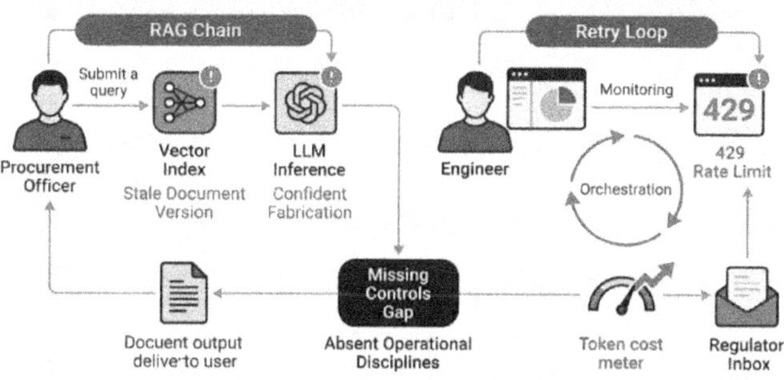

Why This Book, Why Now

Large language models completed one of the fastest transitions from research curiosity to production infrastructure in the history of software engineering. What was a demonstration capability in early 2022 had become a default feature surface by 2024 and a billion-dollar budget line item by 2026. That transition happened faster than the engineering disciplines required to sustain it. Teams that spent two years asking whether to use LLMs are now spending the next two years learning how to operate them with the reliability, cost discipline, safety posture, and observability that the rest of their stack already expects.

The operational gap is real, and it is costing organizations. Prompt injection vulnerabilities have exposed sensitive data in customer-facing assistants. Hallucinated outputs have reached clinical and financial workflows with no downstream correction. Uncontrolled token spend has turned a low-usage feature into a financial escalation. Model behavior has drifted after a provider update with no monitoring signal to catch it. Agentic systems have taken irreversible actions without adequate human-in-the-loop controls. These are the canonical failure shapes of the first generation of production LLM systems, and they share a common cause: teams transferred the fast-moving energy of the prototype phase directly into production without closing the engineering disciplines that separate the two.

Teams that are now operating LLMs in production no longer ask which model to use. They are asking which model, at which size, with which fine-tuning strategy, served by which architecture, monitored by which signals, tested by which gates, and operated within which budget and which compliance regime. Those are concrete engineering questions, and the answers to them have stabilized enough to be taught as patterns. The LLM tooling landscape will rotate again in the next two years, as it has in each of the last four. The patterns will not. A team that has defined its selection framework, serving architecture, safety posture, and observability stack, can evaluate any incoming tool against a stable set of engineering criteria rather than being perpetually disrupted by the next release cycle.

This book exists because the field has matured past the point where the most useful contribution is another survey of model capabilities. The useful contribution now is operational clarity: the patterns, failure modes, decision frameworks, and production disciplines that let a team ship an LLM feature and keep it running. Every chapter is organized around that goal. Every technical checklist is designed to function as a release gate. Every team discussion section is designed to surface misalignments before they surface as incidents.

Who This Book Is For

This book is written for four roles that converge at the same LLM production system: ML architects, ML and LLM engineers, technical team leads, and platform and SRE engineers. Each role brings a different vantage point to the same decisions, and the book is structured to serve all four without forcing any of them to translate the others' vocabulary. Every chapter pairs design-level reasoning with implementation-level detail, and the team discussion sections at the end of each chapter are built for conversations that cross role boundaries.

ML architects will find a framework for making selection and design decisions that age well across platform cycles. Which model family fits this workload's latency and cost envelope? When does fine-tuning add capability that prompt engineering cannot reach? When does a self-hosted deployment justify the operational burden over a managed API? What serving architecture absorbs traffic spikes without cost surprises? The book provides the criteria for those decisions, not just conclusions drawn from a specific configuration. A mission-aligned LLM architecture is one designed from the start to absorb the model updates, provider changes, and traffic growth that production systems inevitably encounter.

ML and LLM engineers will find implementation patterns grounded in the realities of production systems: how to build an evaluation suite that catches quality regression before it reaches users, how to structure a prompt library as managed code rather than a folder of text files, how to instrument an LLM feature so that on-call engineers can localize a quality failure in minutes. The technical checklists at the end of each chapter are designed to be used directly as release gates: specific, actionable, and grounded in the failure modes each chapter names.

Technical team leads will find a shared language to align engineers and architects on reliability standards and a set of discussion guides to surface misalignments before they become production incidents. The pilot-to-production pathway this book describes is one that a team can document, audit, and repeat.

Platform and SRE engineers will find the operational layer of each pattern explicitly named: what the serving infrastructure needs to know about LLM workload shapes, how token budgets interact with cost attribution, which observability signals matter for quality degradation, and where security and compliance boundaries lie in a shared LLM platform. LLM workloads are a distinct workload class. This book treats them as such.

How This Book Is Organized

The twelve chapters are organized into four logical parts, each covering a distinct phase of the LLM production lifecycle. Reading the parts in order gives a complete picture of the discipline from landscape orientation to frontier practice. Reading individual chapters gives actionable guidance on a specific problem without requiring the others as prerequisites. Four appendices provide reference material useful long after the first read: a glossary, a set of reference architectures, release-gate checklists, and compliance mapping templates.

Part I, Foundations (Chapters 1 through 4), establishes the vocabulary and mental models on which the rest of the book depends. Chapter 1 introduces the LLM production landscape: what counts as an LLM, where the deployment models differ, what the canonical business use cases look like, what the LLM-specific risk surface is, and what a working definition of production means for an LLM feature. Chapter 2 covers model selection and sizing: how to compare candidates on capability, latency, and cost, how to run benchmarks that reflect actual workloads rather than public leaderboards, and how to produce a sizing document that the platform team can act on. Chapter 3 covers fine-tuning and adaptation: when instruction tuning, LoRA, PEFT, or preference alignment is the right tool, how to structure a fine-tuning workflow that includes evaluation gates, and how to avoid the common failure modes of domain adaptation. Chapter 4 covers prompt engineering and prompt operations: how to structure prompts for reproducibility, how to manage a prompt library as versioned code, how to evaluate prompt quality systematically, and what a CI/CD pipeline for prompt updates looks like.

Part II, Adaptation and Serving (Chapters 5 and 6), covers the infrastructure that determines whether an LLM feature survives contact with real traffic. Chapter 5 covers serving architectures and latency engineering: the canonical serving topologies, continuous batching, KV cache management, speculative decoding, and the latency optimization patterns that close the gap between model capability and user experience. Chapter 6 covers cost engineering and resource management: token cost attribution, semantic and prompt caching, model routing, autoscaling strategies, and the cost governance patterns that prevent an LLM feature from becoming an uncontrolled budget event.

Part III, Production Concerns (Chapters 7 through 10), covers the disciplines the rest of the engineering organization will hold the LLM team accountable for. Chapter 7 covers safety, security, and compliance: the prompt-injection attack surface, output-safety classification, data-leakage prevention, the EU AI Act and sector-specific regulatory frameworks, and the responsible-by-design patterns that make compliance a build-time property rather than an approval gate. Chapter 8 covers observability and monitoring for LLMs: the signals that standard application monitoring misses, the metric stack that distinguishes quality degradation from infrastructure degradation, trace-level logging for LLM systems, and the alerting patterns that surface LLM failures before users report them. Chapter 9 covers testing, CI/CD, and model validation: the evaluation patterns that make a non-deterministic system deterministic at the release gate; the regression test suite design for LLM features; the canary and shadow deployment patterns for model updates; and the automated evaluation infrastructure that makes continuous deployment safe. Chapter 10 covers human-in-the-loop and feedback systems: the checkpoint placement patterns that insert human oversight without becoming a throughput bottleneck; the feedback collection and label pipeline that feed continuous improvement; and the RLHF and preference alignment workflows that close the loop between user signal and model behavior.

Part IV, Frontier Patterns and Case Studies (Chapters 11 and 12), extends the book's patterns to the current shape of the field and then synthesizes them into reference architectures. Chapter 11 covers retrieval-augmented generation, agentic systems, and tool-using LLMs in production: the RAG pipeline patterns that prevent the stale-document failures described in this introduction, the orchestration and tool-calling architectures that make agents reliable, the loop-prevention and budget-control patterns that prevent the retry explosion also described here, and the evaluation frameworks for multi-step agentic systems. Chapter 12 presents end-to-end case studies across representative industry patterns and synthesizes them into reference architectures that the team can adapt directly. The appendices supply a glossary of terms, detailed reference architecture diagrams, release-gate checklist templates drawn from every chapter, and compliance mapping templates that align workload behavior with major regulatory frameworks.

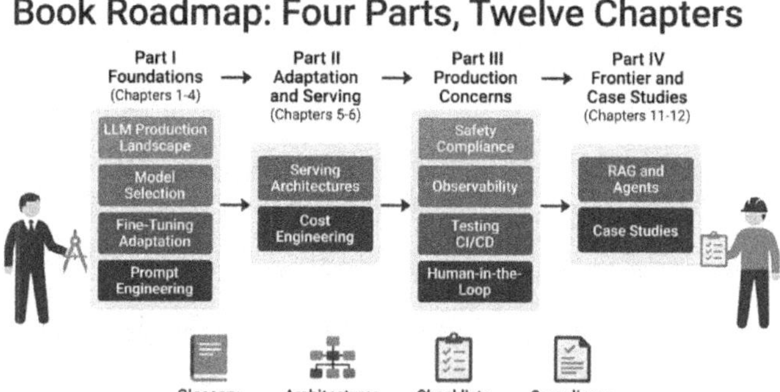

How to Read This Book

Architects and team leads who are new to LLM production infrastructure will benefit most from reading cover to cover, in chapter order. The arc is cumulative. The deployment model vocabulary introduced in Chapter 1 shapes the decisions about the serving architecture in Chapter 5. The fine-tuning patterns in Chapter 3 are prerequisites for the evaluation methodology in Chapter 9. The observability signals defined in Chapter 8 underpin the human-in-the-loop checkpoint designs in Chapter 10. The RAG pipeline patterns in Chapter 11 assume the cost engineering disciplines from Chapter 6 are already in place. Reading sequentially builds the mental model that lets later chapters land with full force rather than as isolated techniques.

ML and LLM engineers with an existing production system and a specific problem will find each chapter usable as a standalone reference. Every chapter opens with a composite scenario drawn from the shapes of real production incidents, positioning the chapter's subject in a recognizable operational context. The core technical content is organized into named patterns with clear criteria for when to apply each. Every chapter closes with a Technical Checklist the team can run as a release gate, a Team Conversation guide for the next architecture review, and a set of key takeaways that distill the chapter's essential argument.

For architects building a new LLM platform from a clean slate, the recommended path is Chapters 1, 2, 5, 6, and 7 in order, followed by Chapter 12 for the reference architectures. This path covers landscape orientation, model selection, serving infrastructure, cost engineering, and safety posture. Chapters 8 and 9 can follow immediately to add observability and CI/CD to the architecture.

For engineers inheriting an existing LLM system tasked with improving reliability, quality, or cost, the recommended path starts with Chapter 8 on observability, Chapter 9 on testing and evaluation, and Chapter 6 on cost control. Those three chapters address the most common operational pain points in mature LLM deployments. They can be read independently of the earlier foundational chapters if the system's architecture is already established.

For platform and SRE engineers supporting LLM workloads they did not design, the recommended entry point is Chapter 1 for the LLM workload vocabulary, Chapter 5 for serving infrastructure, Chapter 7 for the security and compliance surface, and Chapter 8 for observability. Chapter 11 on RAG and agentic systems can be read early as a motivation for the foundational patterns, then revisited after Chapters 5, 6, and 8 have established the serving and observability foundations it builds on.

A Note on Tool Neutrality

The patterns in this book are described independently of any specific platform, framework, or vendor. Model providers, orchestration frameworks, vector databases, serving runtimes, observability platforms, and evaluation tools are referenced as illustrative examples in callouts, not as the spine of the content. The spine is the underlying pattern: the decision that must be made, the failure mode that must be anticipated, and the operational discipline that must be institutionalized, regardless of which tools a team currently uses or will adopt in the next platform cycle.

This choice is deliberate and consequential. The LLM tooling landscape has undergone significant shifts over the past four years. Serving frameworks, orchestration libraries, evaluation platforms, and vector stores that were experimental two years ago are now the de facto standard for certain workload classes. Frameworks that were the de facto standard two years ago have since been superseded. A book organized around specific tools risks becoming a historical document before a second edition is warranted. A book organized around patterns remains useful across platform generations because the patterns describe problems that do not change when the tool names do.

When specific tools are named, they appear to make abstract patterns concrete rather than endorsing a specific vendor or constraining an implementation choice. The same pattern will be implemented differently across organizations and stacks, and the book's value is precisely in naming the pattern above the implementation so the reader can recognize it in any context. Teams should substitute the tools they actually use wherever a callout names a specific example. The pattern is a durable investment. The tool is the current implementation.

Conventions Used in This Book

Each chapter follows a consistent structure. The chapter opens with a composite opening scenario drawn from the shapes of real production incidents, followed by a Why It Matters section that ties the chapter's patterns to mission-level risk and workflow-level impact. The core technical content is organized into named patterns with clear criteria for when each applies, what failure modes it addresses, and what decision it informs. Concepts are tied to the role most responsible for the decision: architect, engineer, lead, or platform partner.

Each chapter closes with three structured artifacts. The Technical Checklist is a specific, actionable list of conditions the team should verify before declaring a system production-ready against the chapter's topic. It is designed to be used as a release gate or architecture review checklist rather than as a summary of the chapter's content. The Team Conversation section provides a discussion guide for the next team meeting or architecture review, with questions designed to surface misalignments among architects, engineers, and platform teams before they surface as production incidents. The Key Takeaways section distills the chapter's essential arguments into a set of positions the team can adopt and defend. The takeaways are stated as forward-facing positions rather than backward-looking summaries.

Diagram placeholders appear throughout each chapter as inline body paragraphs, numbered by chapter and sequence. Diagram 2.3 is the third diagram in Chapter 2. Diagram I.1 is the first diagram in the Introduction. Each placeholder includes a title and a description specific enough for a designer to implement without ambiguity. Diagrams follow a consistent visual style: a blue color palette, a white background, a flat vector look, clean sans-serif typography, and human figures interacting with the systems being described. Red is reserved for errors and failure states only. Diagram placeholders are formatted as body text so they do not appear in the table of contents.

Callouts name a specific tool or framework as an illustrative example of the pattern being described. They should be read as one concrete instance of the pattern, not as a prescription or endorsement. Appendix A provides a glossary organized by domain. Appendix B provides reference architectures for the canonical deployment patterns. Appendix C collects Technical Checklists from every chapter into a single release-readiness reference. Appendix D provides compliance mapping templates aligned with major regulatory frameworks, including the EU AI Act and sector-specific rules for financial services and healthcare.

The Pattern Catalog Framework: Selection to Operations

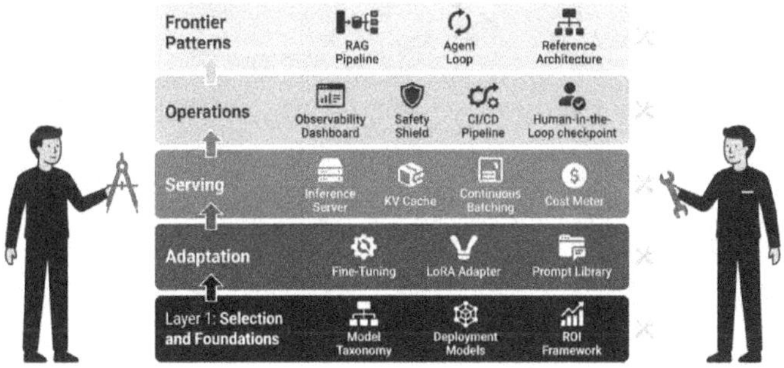

Production LLM engineering rewards operational clarity: teams that have named their patterns, documented their failure modes, and closed their evaluation loops are consistently more reliable, more cost-efficient, and more defensible under compliance review than teams that accumulated their knowledge one incident at a time.

The goal of this book is to compress hard-won LLM production engineering lessons into a pattern catalog that any team can adopt in the order that fits their situation and return to as a reference long after the first read. Each chapter ends where the next begins. The patterns compose. The vocabulary becomes precise enough to make the next incident shorter.

3 About This Series

This book is the fourth volume in the Serious AI Manager Series A, published by Cybersoft Publishing. The series is written for architects, technical leads, and platform engineers who carry production systems into the real world rather than for newcomers building a first prototype. Each volume assumes the discipline established by the volumes before it and extends that discipline into a new layer of the modern AI stack. The progression is deliberate: from the operational foundations of machine learning, through the data and distributed-systems substrate that AI workloads run on, into the specific operational territory of large language models, and finally into the integrated application architecture that ties everything together. The series is a shelf, not a set of independent titles, and it rewards a reader who works through it in order.

3.1 Where This Book Sits

Managing Production Large Language Models occupies the specialization slot in the series. By the point a reader reaches this volume, the assumption is that the foundations from the first three books are already in hand: the MLOps lifecycle and governance model, the production data engineering practices that feed model systems, and the distributed-systems patterns that make large models tractable at scale. This book does not re-teach those foundations. It takes them as given and turns to the operational questions that are specific to large language models, including model selection and sizing, fine-tuning, prompt operations, serving and latency engineering, cost control, safety, observability, testing, human feedback, retrieval-augmented generation, and agentic systems. It is the generative-AI operations volume on a shelf the reader already owns.

3.2 Read These Alongside

The Enterprise MLOps Playbook: How to Deploy, Govern, and Scale Machine Learning Systems in the Real World — Jazper Carter. Read this first for the lifecycle, governance, and deployment foundations that the present book assumes throughout.

Production Data Engineering for Machine Learning — Claude J. Louis-Charles, PhD. Read this for the data pipelines, feature management, and quality controls that feed every model system, including LLM retrieval corpora.

Distributed Machine Learning Patterns: A Patterns-First Manual for Architects, Engineers, and Technical Leads — Jazper Carter. Read this for the architect-view treatment of distributed serving, sharding, and large-model patterns referenced here at operational depth.

Designing Data-Intensive AI Applications: An Architect's Blueprint for AI Systems Under Real-World Scale, Latency, and Compliance Pressure. Read this last for the application-architect synthesis that closes the series.

Taken together, the five volumes move from foundations to specialization to synthesis. This fourth book is the point on that arc where the general discipline of operating machine learning becomes the specific discipline of operating language models, and it assumes that a reader arriving here has already absorbed the lifecycle, data, and distributed-systems material rather than encountering it for the first time. The boundaries between volumes are deliberate: where this book references distributed serving or large-model partitioning at operating depth, it points to the third volume for the design-time treatment rather than reproducing it. The fifth and final volume closes the series with the application-architect view, integrating the operational practices established here into end-to-end system design under scale, latency, and compliance pressure.

4 Introduction and the LLM Production Landscape

4.1 Opening Scenario

It is 9:47 on a Wednesday morning when the product lead for a mid-sized enterprise's new assistant feature receives a calendar invite titled "LLM Launch Alignment." The night before, a fifteen-minute demo to the executive team had gone better than anyone anticipated. The lead VP of product had asked the room for a ship date. The head of engineering, who had been watching the demo from the back of the conference room, said, "End of quarter." A graphic designer in the front row had already started sketching the announcement post.

By Thursday afternoon, the product lead has assembled the core team: a senior ML engineer who built the prototype, a backend engineer who wired it to the existing API layer, and a product manager who owns the feature roadmap. The four of them sit down to scope the work. The prototype is a single Python function that calls a third-party model API, passes a system prompt, and returns the completion. It took two days to build. It works beautifully in the demo environment, where the inputs are curated, the session is single-user, and the cost of a wrong answer is zero.

The scoping session raises 12 questions that the team cannot answer. What happens when three thousand users call the feature simultaneously? Which model do they use at scale, and what does it cost per request at that volume? What is the latency budget the mobile client can tolerate? What does the safety posture look like when the feature is exposed to users who are not product managers? What happens when the upstream model provider has an outage? Has legal reviewed the data-handling implications of sending user inputs to a third-party inference endpoint? Does the workload fall under any sector regulation that requires logging, auditability, or human review? How does the team know when the model's answer quality degrades after a provider-side model update? What does rollback look like? Who owns the on-call rotation for a feature whose failure mode is not an error code but a plausible-sounding wrong answer?

None of these questions invalidates the prototype. The prototype proved that the use case is real, that users find it valuable, and that the core model capability exists. But the gap between what the team demonstrated and what they need to ship is the entire operational surface of a production LLM system. The gap is not a sign that the team is behind. It is a sign that they are at the beginning of a well-defined engineering problem, one that the field has now accumulated enough operational experience to teach as patterns rather than as improvised solutions.

This book answers those twelve questions. This chapter provides the map: the taxonomy of models the team will choose from, the deployment models that shape every infrastructure decision, the business and ROI framework that gives the work its acceptance criteria, the risk surface the team is now responsible for, the regulatory context that compliance and legal will raise, the metric stack that distinguishes a working prototype from a production capability, the build-buy-compose decision the team will revisit repeatedly, the roles that own each layer of the system, and the working definition of the word 'production' that every later chapter tests against.

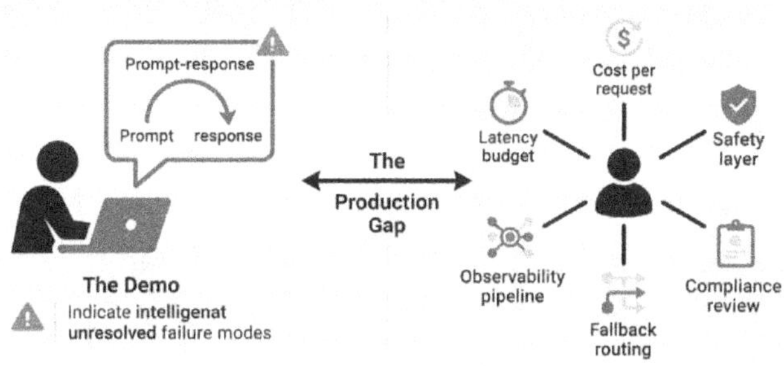

4.2 Why It Matters

Large language models are now a default feature surface for software teams across virtually every sector. The trajectory from research curiosity to production infrastructure happened faster than any comparable technology in the previous decade. The teams building on these models inherited a new operational surface with no established playbook, a cost structure unlike anything in their existing architecture, and failure modes that do not trigger the alarms they already have. A model that silently begins returning lower-quality answers after a provider-side update looks identical to a healthy model on a traditional infrastructure dashboard. A prompt injection attack that causes the assistant to reveal system instructions does not show up as elevated error rates. These are production problems that require production discipline.

The workflow-level impact of getting this wrong compounds quickly. A team that ships an LLM feature without a defined latency budget discovers the problem when the mobile client times out in the first week of real traffic. A team that ships without a safety posture discovers the problem when a screenshot of a harmful completion reaches the company's social feed. A team that ships without a cost model discovers the problem when the monthly infrastructure invoice arrives, revealing that the feature's gross margin is negative. These are recoverable situations, but each recovery costs weeks of engineering time that could have been spent shipping the next capability.

The mission of this book is operational clarity for teams that have already decided to ship LLM features. The orientation in this chapter is not about whether to use large language models. That question is settled for the audience this book addresses. The questions on the table are which model at which size, with which adaptation strategy, served by which architecture, monitored by which signals, evaluated by which gates, and operated within which budget and compliance regime. This chapter frames all those questions as tractable engineering decisions, each with a pattern, an owner, and a measurable definition of done.

4.3 What Counts as an LLM: A Taxonomy of Models

The term 'large language model' is used loosely in industry conversations, often applied to any transformer-based text model, regardless of size, capabilities, or architectural variant. For engineering purposes, precision matters. The model a team chooses determines its cost per token, latency profile, fine-tuning surface, compliance footprint, and deployment options. A team that treats all models as interchangeable substitutes will make poor decisions at every downstream decision point. A shared taxonomy is the first operational tool.

Small language models occupy the lower end of the parameter range, typically from one billion to around twenty billion parameters. They are fast, inexpensive to serve, can often run on a single GPU or even on-device hardware, and are highly amenable to task-specific fine-tuning. Their capability ceiling is real: they handle focused, well-defined tasks with high fidelity but struggle with open-ended, multi-step reasoning, broad world-knowledge recall, and complex instruction-following. For structured extraction, classification, short-form generation, and narrow-scope domain-specific assistants, small models are often the mission-aligned choice, with substantial cost and latency advantages.

Frontier large language models sit at the upper end of the public capability curve at any given point in time. They are trained on broad corpora at scale, exhibit strong generalization across tasks, and handle complex instruction following, nuanced reasoning, and long-context synthesis with a quality ceiling that smaller models cannot match. They are also the most expensive to call, the most latency-sensitive when run synchronously, and the most complex to evaluate because their output space is broad. Teams that reach for a frontier model by default, without asking whether their task actually requires frontier capability, will find their cost model unsustainable at production volume.

Reasoning models are a distinct architectural variant that has matured significantly since 2024. They are optimized for multi-step problem-solving, produce chain-of-thought traces as a core output artifact, and are designed for tasks where the correctness of a derivation matters more than fluency or speed. Their per-request latency is higher than that of frontier general-purpose models because the reasoning trace is generated token-by-token before the final answer is produced. For code generation with correctness requirements, mathematical reasoning, and structured decision support, reasoning models offer a quality profile that general-purpose frontier models do not.

Multimodal models accept input modalities beyond text, including images, audio, and, in some variants, video and documents. From a production engineering perspective, they introduce additional preprocessing pipelines, larger per-request payloads, new categories of content-moderation requirements, and, in some jurisdictions, additional regulatory obligations related to biometric or sensitive visual data. A team building a multimodal feature should treat the modality handling as a distinct engineering workstream from the language model integration.

Fine-tuned specialized variants are base models that have been adapted, through supervised fine-tuning, reinforcement learning from human feedback, or parameter-efficient methods, to a specific domain, task, format, or persona. They appear on the taxonomy as a derived layer rather than a base category: any of the model classes above can produce a fine-tuned variant. From a production standpoint, a fine-tuned model introduces a versioned artifact with its own evaluation gate, rollback path, and training data governance obligations. The adaptation strategy a team chooses is the subject of later chapters; what matters here is that fine-tuning is a production operation, not a one-time setup step.

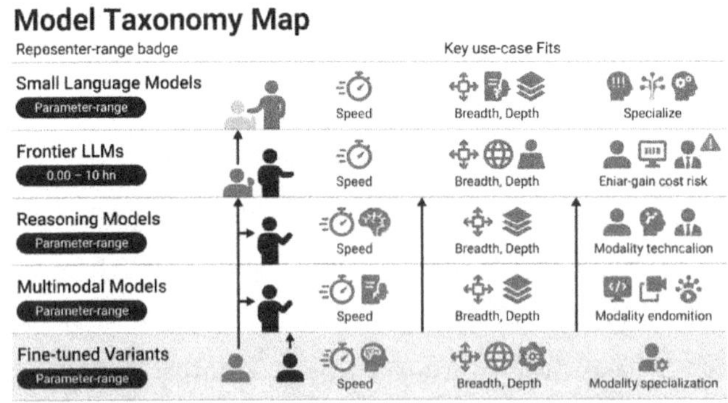

The taxonomy is not a ranking of quality or prestige. A small model that is well-matched to a task and well-operated in production is a better engineering outcome than a frontier model operated carelessly. The engineer's discipline is to locate the candidate model on the taxonomy, state the reasons the task requires that category, and reject or defer the others with documented reasoning. That documentation is the first artifact requested by the Technical Checklist at the end of this chapter.

4.4 On-Premises, Cloud, Hybrid, and Edge Deployment Models

The deployment model is the second foundational decision, and it shapes every infrastructure, cost, latency, and compliance decision that follows. There are four canonical deployment models for production LLMs: cloud-managed, self-hosted (which includes on-premises and private cloud), hybrid, and edge. Teams that choose a deployment model by default rather than by deliberate evaluation against their workload's constraints will find themselves locked into architectural compromises they cannot easily undo.

Cloud-managed deployment means calling a model through a vendor's inference API endpoint. The operational burden is low: the vendor manages hardware provisioning, model updates, scaling, and availability. The cost structure is consumption-based and predictable at low volume, but can become a dominant line item at high volume or with long-context workloads. Data residency is governed by the vendor's terms and region selection, which may not meet the workload's data handling obligations. The vendor controls model versioning; a provider-side model update can change output behavior without advance notice unless the team pins to a specific model version and the vendor supports pinning. For teams early on the pilot-to-production pathway, cloud-managed is the lowest-friction starting point, but it is not a production architecture on its own.

Self-hosted deployment means running the model on infrastructure the team controls, whether in an on-premises data center or a private cloud. The operational burden is substantially higher: the team is responsible for GPU provisioning, serving framework configuration, model artifact management, hardware maintenance, and availability. The cost structure shifts from per-token consumption to infrastructure fixed cost plus utilization, which can become dramatically more efficient at high sustained request volume. Data residency is fully controlled, which satisfies strict data sovereignty requirements. Model versioning is also fully controlled, enabling reproducible evaluation gates and predictable rollback. Self-hosted deployment requires a platform partner with GPU operations competency that many teams do not yet have when they begin their LLM journey.

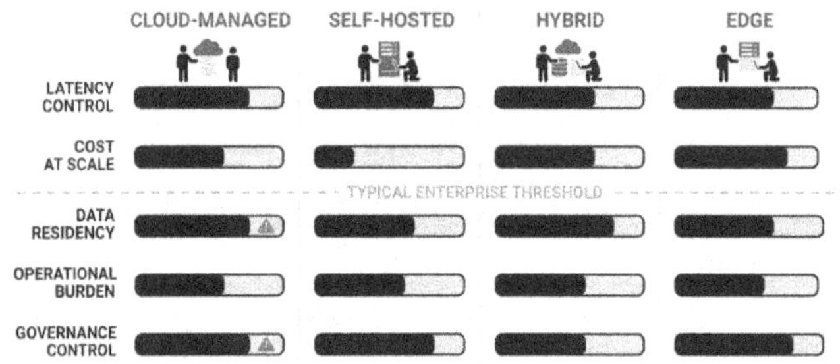

Hybrid deployment combines a cloud-managed endpoint for baseline traffic with a self-hosted endpoint for high-volume or sensitive workloads. The routing layer that sits between the application and the model backends is the new engineering surface this model introduces: the team must decide which requests go where, how to handle failover, and how to ensure consistent output behavior across two model sources that may not be identical. Hybrid architectures are the most operationally complex of the four models. Still, they are often the right long-term solution for teams with both variable-traffic features and data-sensitive workloads.

Edge deployment runs the model on a device or at a network edge node rather than in a centralized data center. It is the correct choice when network round-trip latency is prohibitive, when the device must operate offline, or when data must not leave the device. Edge deployment is currently constrained to small models because of device memory and compute limits. However, quantization and distillation techniques have meaningfully expanded what is possible at the edge in recent years. The model update lifecycle for edge-deployed models introduces a device fleet management problem that most ML teams have not previously operated.

The decision rubric that the team should apply maps the workload's constraints to these four models. The constraints are: the latency budget the application can tolerate; the expected request volume and its variability; the data classification of the inputs and outputs; the regulatory regime's data residency requirements; and the team's operational maturity for GPU infrastructure. A workload with a strict latency budget, moderate volume, and no data residency requirements maps naturally to cloud-managed. A workload with high sustained volume and strict data residency maps to self-hosted. A workload with a mix of both maps to hybrid. A workload with offline or on-device requirements maps to the edge. The rubric is not a formula; it is a structured checklist that forces the team to name its constraints before it picks its architecture.

4.5 Business Use Cases and an ROI Framework

Not all LLM use cases are equal in their return on investment, their operational complexity, or their risk surface. A team that cannot articulate the business case for its LLM feature in terms a finance partner can review will find the conversation difficult when the monthly inference bill arrives. The field has now accumulated enough production experience to describe the categories of LLM use cases that have proven economically viable, and to frame each with an ROI structure that survives the first finance review.

The assistant pattern applies to conversational interfaces in which the model answers questions, guides workflows, or provides information retrieval support within a bounded domain. The business value driver is deflection: the assistant handles a volume of requests that would otherwise require a human agent, a support ticket, or a search session that ends without resolution. The ROI calculation requires a baseline deflection rate, per-interaction cost estimates for both the human and model sides, and a quality threshold below which deflected interactions become re-escalations that cost more than the original interaction.

The summarizer pattern covers workloads in which the model reduces a long input artifact, such as a document, meeting transcript, contract, or email thread, into a shorter, structured output. The business value driver is time recovery: the model returns minutes per worker per day that compound across a large knowledge workforce. The ROI calculation requires an estimate of the average time saved per summarization event, an assumption about the number of events per worker per week, and a loaded hourly rate for the worker population. Summarization is among the lowest-risk LLM patterns in production because the source document is always available for human review, and the cost of an imperfect summary is a re-read rather than a consequential error.

LLM Use Case Taxonomy and ROI Drivers

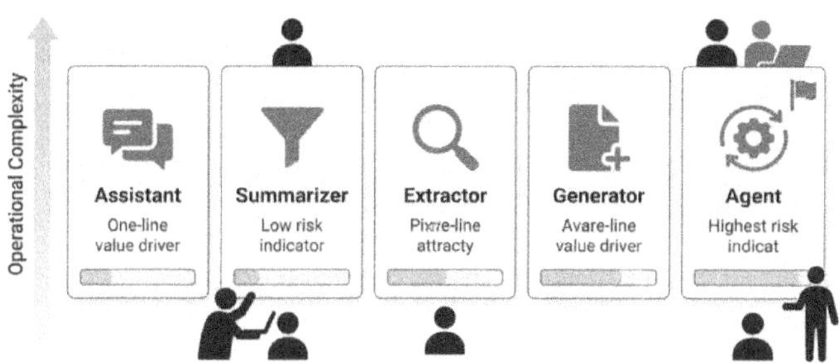

The extractor pattern addresses workloads in which the model identifies and outputs structured data from unstructured inputs, such as named entities from documents, line items from invoices, key clauses from contracts, or structured fields from free-text forms. The business value driver is automation of a previously manual, labor-intensive data entry or review task. The ROI calculation requires a per-record processing-cost baseline for the human process and an accuracy threshold below which the automated output must be re-reviewed, thereby eroding the automation gain. Extraction workloads are often well-suited to smaller, fine-tuned models because the output schema is fixed and the model's task is precisely bounded.

The generator pattern covers workloads where the model produces original content from a specification: marketing copy, code, test cases, data augmentation, or draft documents. The business value driver varies significantly by workload; code generation has the clearest ROI signal because the output is objectively evaluable by a compiler or test suite. Content generation ROI is harder to measure and depends on downstream conversion or engagement metrics that are slower to validate. Generator patterns carry a higher editorial and brand-risk surface area than summarizer or extractor patterns and typically require a human-review gate in the workflow.

The agent pattern covers workloads in which the model orchestrates multi-step tasks by calling external tools, reading intermediate results, and deciding on the next action in a loop. Agents represent the highest level of operational complexity among the five patterns: the failure surface includes tool-call errors, reasoning loops, unauthorized action scope, and compounding errors across steps. The ROI potential is also the highest, because agents can automate workflows that previously required a human to manage multiple systems in sequence. Responsible-by-design agent architecture requires explicit scope boundaries, tool call auditing, and a human-in-the-loop review gate for consequential actions. Chapter twelve covers agent patterns in production in depth.

A production ROI framework for LLM features has three layers. The first layer is the quality baseline: the team must establish what 'good enough' means for the model's output before any economic calculation is valid, because a model that is cheap but wrong is not cheaper than a human. The second layer is the cost model: total cost per unit of value delivered, which includes inference costs, human review costs, engineering and maintenance costs, and the amortized cost of evaluation infrastructure. The third layer is the business metric: the lagging indicator the business actually cares about, whether that is deflection rate, time-to-resolution, error rate reduction, or revenue per session. A team that can articulate all three layers has a finance-reviewable business case. A team that can articulate only the technical cost has a tool, not a product.

4.6 Risks, Failure Modes, and the New Operational Surface

Every production system has a risk surface. LLMs introduce a risk surface that most engineering teams have not previously operated on, with failure modes that do not match the failures their existing monitoring is designed to detect. Understanding these failure modes at the start of the project, rather than at the first incident, is the difference between a team that has planned for failure and a team that is surprised by it.

Hallucination is a class of failures in which the model generates output that is factually incorrect, internally inconsistent, or unsupported by any source it had access to, while presenting that output with the same fluency and confidence as a correct answer. From an operational standpoint, hallucination is dangerous because it is invisible to standard monitoring: the API returns a 200, the response is well-formed, the latency is within budget, and the content is wrong. Detection requires an evaluation layer that scores output quality against a ground truth or a reference corpus, and that layer must run continuously in production, not only before deployment. Chapter nine covers hallucination detection patterns.

Prompt injection is a class of attack in which a malicious payload in the user's input, or in content the model retrieves from an external source, causes the model to override its system instructions, reveal confidential context, or take unintended actions. Unlike a traditional injection attack, prompt injection does not exploit a code vulnerability; it exploits the model's instruction-following capability. Defenses against prompt injection are architectural: input sanitization, system prompt isolation, output filtering, and, in agentic systems, explicit scope constraints on what actions the model is permitted to initiate. Chapter seven covers prompt injection defense in depth.

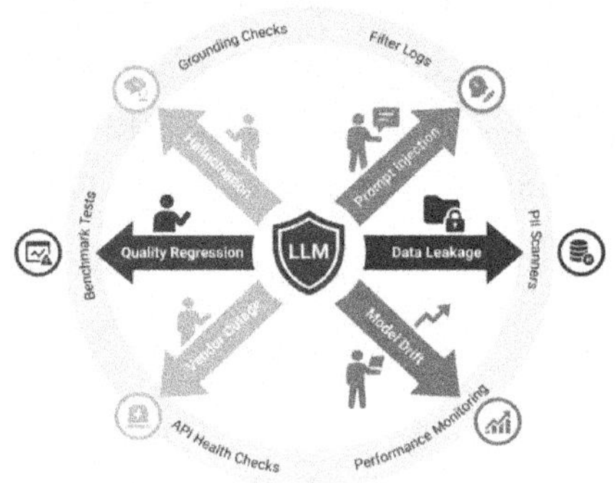

Data leakage is a class of failures in which the model reveals information that should be confidential: a user's data to another user, a system prompt containing proprietary instructions, or a retrieved document that the requesting user is not authorized to view. Data leakage can occur through the model's context window, through insufficiently isolated retrieval results, or through a conversation memory system that persists across user sessions without proper tenant isolation. The engineering controls are session isolation, retrieval-layer authorization checks, and output scanning for patterns that match sensitive data formats.

Model drift is the class of failure where the model's output behavior changes over time, either because the production model was silently updated by the provider, because the distribution of real-world inputs has shifted away from the evaluation distribution, or because a fine-tuned model has been retrained on new data that shifted its behavior in an unintended direction.

Model drift is particularly insidious because the change is gradual and often visible only in aggregate quality metrics over days or weeks, not in individual request failures. Continuous evaluation against a stable benchmark set is the primary detection mechanism.

Vendor outage is not an LLM-specific risk, but it takes a different shape for teams that depend on a cloud-managed model API as a core product dependency. Traditional dependency outages degrade a feature; an LLM API outage can turn off the entire feature surface if no fallback architecture exists. The engineering response is a defined fallback strategy: a secondary model endpoint, a graceful degradation path that switches to a smaller local model, or a queuing architecture that holds requests during short outages and drains when the primary endpoint recovers. The fallback strategy must be tested before the outage occurs, not designed during it.

Quality regression is a class of failures in which a change to the system, whether a prompt update, a model version change, a retrieval corpus update, or a fine-tuning run, degrades the quality of the model's outputs without triggering any infrastructure alert. Quality regression is the LLM-specific analog of a software regression, and it requires the same engineering response: a regression test suite, a defined quality gate, and a CI/CD pipeline that runs the evaluation suite before any change reaches production. Chapter eleven covers LLM CI/CD and evaluation gates in detail.

4.7 Regulatory Context: EU AI Act, Sector Rules, and Internal Policy

The regulatory landscape for AI systems has moved from a future concern to a present constraint. Teams that treat compliance as a late-stage approval gate rather than a design input will find themselves rebuilding architecture to satisfy requirements they could have accommodated cheaply at the start. The responsible-by-design principle applies directly here: known regulatory requirements are engineering inputs to the system design, not audit findings after the fact.

The EU AI Act is the most comprehensive horizontal AI regulation currently in force. It applies to any team whose product is offered to users in the European Union, regardless of the team's headquarters. Its risk-based framework classifies AI systems into four tiers: unacceptable risk (prohibited), high risk (subject to conformity assessment, documentation, and human oversight requirements), limited risk (subject to transparency obligations), and minimal risk (largely unregulated). LLM-powered features most commonly fall into the limited- or high-risk tiers, depending on the domain. A customer-facing assistant for a general consumer product typically falls into the low-risk tier and must disclose that the user is interacting with an AI system. An LLM used to make or inform decisions about credit, employment, education, or access to essential services falls into the high-risk tier and requires a conformity assessment, bias testing, logging, and a human oversight mechanism.

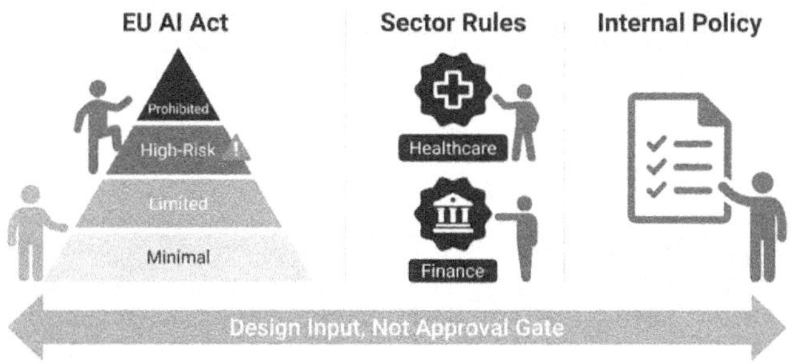

Sector-specific regulation adds an overlay to the horizontal framework. Teams building LLM features in financial services must consider whether their use case falls under existing model risk management guidance, which, in several jurisdictions, requires documentation of the model methodology, validation against independent benchmarks, and ongoing performance monitoring by a risk function separate from the development team. Teams in healthcare must evaluate whether their feature constitutes a regulated clinical decision support tool, triggering obligations for clinical validation, adverse event reporting, and labeling. Teams in regulated industries who are unsure whether their LLM feature is a regulated system should make that determination before the architecture is fixed, not after the system is live.

Internal AI-use policies have become standard in organizations of any significant scale. These policies typically define which model vendors are approved for use, which data classifications are permitted to leave the organization's perimeter, which use cases require a formal risk assessment before deployment, and what human oversight mechanisms must be present for consequential AI-assisted decisions. A team that begins building on a model vendor that has not undergone the organization's security and vendor review process will face a costly, schedule-consuming rework cycle later. The first questions the architect should ask when scoping an LLM feature are: which vendors are on the approved list, what data classification applies to the inputs, and whether this use case requires a formal risk assessment.

Positioning compliance as a design input rather than a late-stage gate is not only a risk-management posture; it is also a practical engineering advantage. A system designed with logging from the start is cheaper to audit than one that needs a logging retrofit. A system designed with human oversight hooks in the workflow is easier to validate than one that requires a governance layer to be bolted on. A system designed with data residency constraints in mind from the beginning does not require a deployment model change six months after launch. Regulatory requirements, when incorporated early, are features of the architecture. When incorporated late, they are considered technical debt.

4.8 Success Metrics and KPIs for Production LLM Projects

A production LLM feature with no defined metrics is not in production; it is in a permanent beta. The metric stack that distinguishes a working prototype from a production capability has three layers: quality metrics, operational metrics, and business metrics. Each layer has distinct ownership, measurement cadence, and consequences when it degrades. A team that conflates these layers, or that measures only the layer easiest to instrument, will have blind spots in its production posture.

Quality metrics assess whether the model produces correct, useful, and safe outputs. The primary quality dimensions are correctness (factual accuracy against a verifiable ground truth, where applicable), faithfulness (the degree to which the output accurately represents the source context, critical for retrieval-augmented systems), helpfulness (whether the output addresses the user's actual intent), and safety (the absence of harmful, toxic, or policy-violating content). Quality metrics are owned by the LLM engineer or the quality function on the team, measured through a continuous evaluation pipeline, and reviewed at a cadence that matches the system's rate of change.

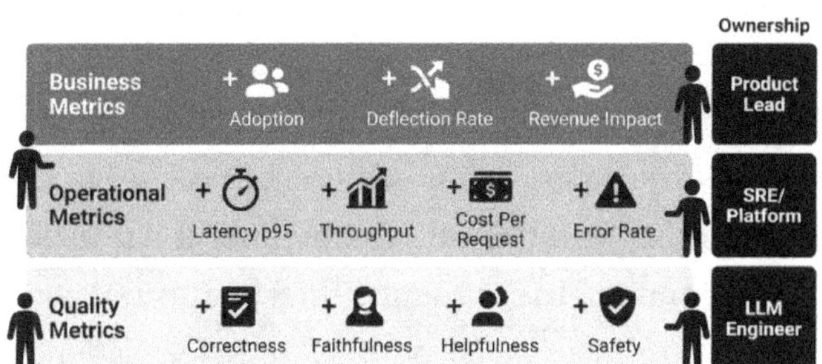

Operational metrics measure whether the system is serving requests reliably and within cost constraints. The critical operational metrics for an LLM feature are latency at the 50th, 95th, and 99th percentiles (because LLM latency distributions are often right-skewed and mean latency can be misleading), throughput in requests per second or tokens per second, cost per request broken down by model inference cost and any downstream processing cost, and error rate including both hard errors and soft failures such as rate-limit responses and truncated outputs. Operational metrics are owned by the platform or SRE partner, measured by infrastructure monitoring, and reviewed in the standard operational review cadence.

Business metrics are the lagging indicators that the organization's leadership cares about. For an assistant feature, the primary business metric is typically the deflection rate: the fraction of interactions in which the model resolves the user's need without escalating to a human. For a productivity tool, the metric is time saved per user per week. For a content generation tool, the metric is conversion rate or engagement on generated content relative to human-authored content. The product lead owns business metrics measured through product analytics and reviewed as part of the business review cadence. A team that can show a healthy metric stack across all three layers has an unambiguous answer to whether the feature is working.

Each metric must have a named owner and a defined threshold that triggers a review. A quality metric without an owner is not a metric; it is a number that will be noticed only after it has already degraded. An operational metric without a threshold is not a gate; it is an observation. A business metric without a review cadence is not a success criterion; it is a dashboard that no one looks at. The metric stack document is a living artifact that the team updates when the system changes, when the business goals change, and when production experience reveals that a metric they were measuring was not capturing what they thought it was.

4.9 The Build, Buy, and Compose Decision

Every LLM project encounters a build-buy-compose decision, and most encounter it multiple times as the system evolves. The decision is not one-time: as model capabilities improve, the team's operational maturity grows, and the feature's cost structure clarifies, the right answer at the six-month mark may differ from the right answer at launch. A team that makes this decision once and never revisits it is likely operating a suboptimal architecture within a year.

Using a managed API is the default starting point for most teams and the right choice for many production workloads. It minimizes time to production, eliminates the operational burden of GPU infrastructure, and provides access to the latest model capabilities without requiring training or serving investment. The costs include consumption-based pricing at scale, vendor-imposed data residency constraints on the vendor's infrastructure, and model behavior that can change with provider-side updates. Teams that choose managed API should define which model version they are pinning to, what their fallback endpoint is, and at what volume threshold they will evaluate self-hosting.

Self-hosting becomes attractive when the combination of volume, data residency requirements, and customization needs tips the cost-benefit calculation. The infrastructure investment required to self-host a large model is substantial: GPU allocation, serving framework expertise, model artifact storage and versioning, and the operational overhead of a GPU fleet are all new capabilities for most teams. The efficiency gains at high sustained volume can be significant, and the elimination of data residency constraints is frequently the deciding factor for teams in regulated industries. Self-hosting a model that is not yet delivering business value is a premature optimization. Self-hosting a model that is already generating a profitable unit economics profile is a sound engineering investment.

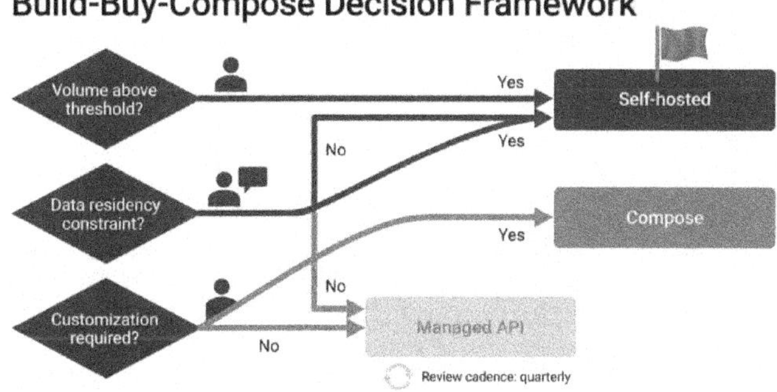

Fine-tuning is a form of build: the team invests in creating a model artifact tailored to their task, domain, or format requirements. Fine-tuning is warranted when the base model's general capability does not meet the quality bar for the specific task, when the task has a narrow and well-defined output format that can be taught more efficiently through examples than through prompting, or when the team needs to bake in domain knowledge that is not well-represented in the base model's training data. Fine-tuning is not warranted as a first response to quality problems that have not yet been diagnosed; prompt engineering and retrieval augmentation should be exhausted first because they are faster to iterate and cheaper to maintain.

Composing multiple models into a single feature means using different models for different steps in the processing pipeline: a small, fast model for initial classification or intent detection, a larger model for generation, a specialized model for a particular format or domain, and potentially a separate model for output evaluation. Composition increases the overall system's flexibility and efficiency. Still, it adds routing logic, additional latency from serial model calls, and more complex evaluation requirements because the quality of the composed output depends on the behavior of multiple models in sequence. Composition patterns are a mature area of production LLM architecture and are covered in depth in later chapters.

The build-buy-compose decision should be reviewed on a defined cadence, not revisited only when a crisis forces it. A quarterly review that asks whether the current architecture still aligns with the volume, cost, and quality profile is a lightweight governance mechanism that prevents architectural drift. The review should be a standing agenda item for the technical lead, with input from the platform partner on cost and operational trends and from the LLM engineer on quality and capability trends.

4.10 LLM Project Roles: Architect, Engineer, Lead, and Platform

LLM production systems require a small set of distinct roles with clear ownership boundaries. Ambiguity about who owns which layer of the system is a leading predictor of incidents that go undetected, of metric gaps that persist without remediation, and of architecture decisions that get deferred until they are expensive. Defining these roles at the start of the project is not organizational overhead; it is a technical prerequisite for a system that can be operated responsibly over time.

The ML architect owns the system design: the model selection decision, the deployment architecture, the integration patterns between the LLM layer and the rest of the application, the evaluation framework, and the long-term technical direction of the LLM platform. The architect is accountable for the build-buy-compose decision and its periodic review, for the selection of serving infrastructure, and for ensuring that the system's design is consistent with the organization's regulatory and security requirements. In smaller teams, the architect role may be filled by a senior engineer rather than a dedicated position, but the responsibilities must still be explicitly owned.

The LLM engineer owns the model-layer implementation: prompt engineering, the fine-tuning pipeline (where applicable), the retrieval system (where applicable), the evaluation pipeline, and the quality metrics. The LLM engineer is the primary owner of the quality metric stack and is responsible for the evaluation gate in the CI/CD pipeline. They are also the primary contact for model behavior questions: why the model is producing a particular class of output, whether a quality regression is attributable to a prompt change or a model update, and whether a proposed prompt or fine-tuning change will improve quality without introducing new failure modes.

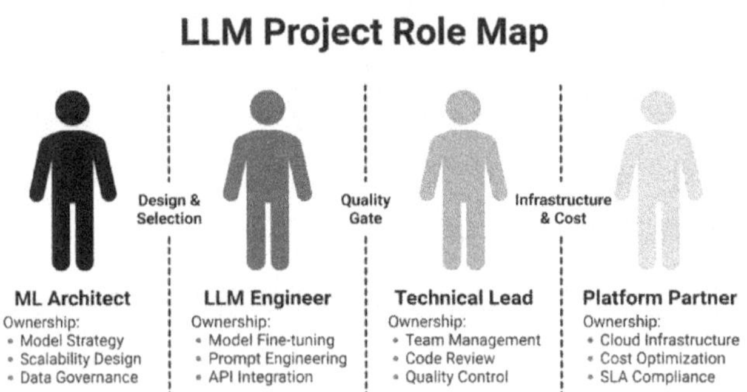

The technical team lead owns the workflow-level integration: product requirements, sprint planning, stakeholder communication, incident coordination, and business metric review. The lead is responsible for ensuring that the metric stack is documented with named owners, that the compliance and legal review is completed before launch, and that the on-call rotation and incident response procedures are defined. The lead is also responsible for the build-buy-compose review cadence and for ensuring that the team's architecture decisions are documented with the reasoning that drove them.

The platform partner owns the infrastructure layer: GPU provisioning, serving framework configuration, availability monitoring, cost allocation, and the network and security controls between the LLM layer and the rest of the application. The platform partner is the primary owner of operational metrics and the first escalation point for infrastructure-layer incidents. In organizations with a shared ML platform team, the platform partner may serve multiple LLM product teams simultaneously; the seam between the product and platform teams must be explicitly defined to avoid gaps in ownership.

The seams between these roles are where integration failures occur if not managed deliberately. The seam between the architect and the LLM engineer is the model selection and evaluation framework: both must agree on what the evaluation criteria are and how they are measured. The seam between the LLM engineer and the platform partner is the serving infrastructure and the quality-to-availability tradeoff: the engineer knows what latency the model requires, and the platform partner knows what the infrastructure can deliver. The seam between the lead and all other roles is the production definition itself, covered in the final section of this chapter.

4.11 How to Read This Book and the Companion Repository Overview

This book is structured to support both linear and non-linear reading. The chapter sequence follows the lifecycle of an LLM feature in production: landscape and orientation (chapters one through three), model selection and adaptation (chapters four through six), serving and cost engineering (chapters seven and eight), safety and security (chapters nine and ten), observability and evaluation (chapters eleven and twelve), CI/CD and human-in-the-loop systems (chapters thirteen and fourteen), and specialized patterns in RAG, agentic systems, and end-to-end case studies (chapters fifteen through seventeen). A reader building a new LLM feature from scratch will benefit from reading chapters one through eight in order. A reader who has an existing feature in production and is responding to a specific operational problem will benefit from reading the chapter that addresses that problem directly.

Each chapter follows the same template. The opening scenario is a composite incident drawn from real operational patterns, presented as a concrete narrative that grounds the chapter's technical content in a recognizable situation. The core content covers eight to eleven canonical patterns, each named and described with the precision an engineer needs to recognize the pattern in their own system and apply it. The technical checklist at the end of each chapter is a production-readiness gate that the team can use before declaring the chapter's topic addressed. The team conversation section provides discussion questions a technical lead can use to pressure-test the team's alignment on the chapter's topic. The key takeaway section synthesizes the chapter's content into a set of durable principles.

The companion repository contains reference implementations for the patterns covered in each chapter, organized by chapter number. Each reference implementation is written to be framework-neutral where possible, with notes indicating where a framework-specific version is available. The repository also contains the evaluation harnesses referenced in the observability and CI/CD chapters, the prompt templates referenced in the prompt engineering chapters, and the checklist templates that correspond to each chapter's technical checklist. The repository is a starting point and a reference, not a production-ready system; every implementation will require adaptation to the specific constraints of the team's environment.

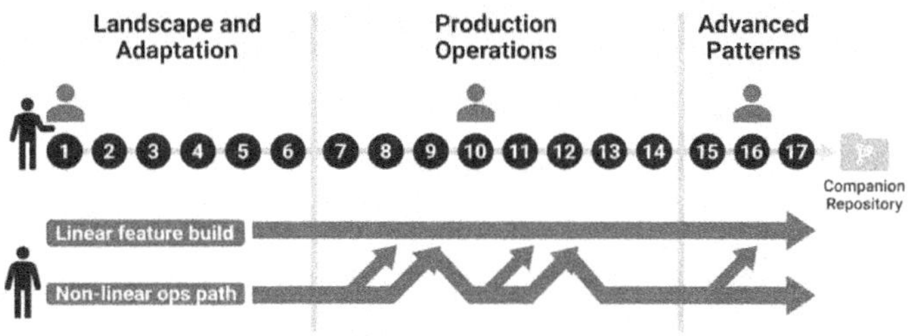

Non-linear reading is encouraged. A reader who is responsible for the safety and security posture of an existing LLM feature can read chapters nine and ten without having read the model selection chapters. A reader who is setting up an evaluation pipeline can read chapters eleven and twelve directly. The terminology and role definitions established in this chapter are referenced throughout the book; readers who jump into later chapters may find it useful to scan this chapter's taxonomy, deployment model rubric, and role map before proceeding.

4.12 A Working Definition of 'Production' for LLMs

The word 'production' is used loosely in software engineering to mean any system that is live and serving real users. For LLM features, that definition is insufficient. A system can be live and serving real users while lacking an evaluation gate, a defined latency budget, a safety posture, a cost model, and a compliance review. That system is not in production in any meaningful operational sense; it is a demo with traffic. The discipline of production LLM engineering requires a definition of 'production' that is specific enough to be tested, measured, and enforced as an acceptance criterion.

A working definition of production for an LLM feature has six components. Reliability means the feature has a defined availability target, a tested fallback path for model endpoint failures, and an on-call rotation with a documented escalation path. Observability means all three layers of the metric stack are instrumented, each metric has a named owner, and there is an alerting threshold that triggers a review before the metric degrades to a user-visible level. Safety means the feature has a documented safety posture covering the failure modes described in this chapter, with specific controls named and verified, and a mechanism for monitoring safety compliance in production. Cost discipline means a cost model exists, the per-request cost is known and reviewed against a budget, and there is a defined threshold at which the cost model triggers an architectural review.

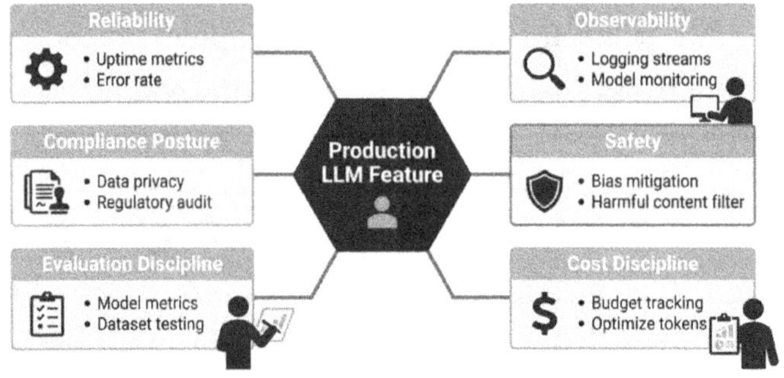

Evaluation discipline means the feature has a regression test suite that runs in CI, a defined quality gate that any change must pass before reaching production, and a continuous evaluation pipeline that monitors production quality against a stable benchmark. A defensible compliance posture means the feature has been reviewed against the applicable regulatory regime, any required documentation has been completed, the data handling has been reviewed against internal policy, and a review cadence has been set to keep compliance current as the system evolves and as the regulatory environment changes.

This six-component definition is the implicit acceptance criterion for every chapter that follows. Each chapter addresses one or more components in depth and concludes with a checklist the team can use to verify that the component is addressed. A team that completes all the checklists in this book has not guaranteed a perfect system; no definition of production guarantees perfection. But a team that completes the checklists has a system with documented reliability targets, observable metrics with named owners, a tested safety posture, a finance-reviewed cost model, a continuous evaluation pipeline, and a current compliance review. That team has a production-grade LLM feature. Everything else in this book is the path to get there.

The pilot-to-production pathway is the journey from a demo that works in a controlled setting to a system that satisfies all six components. The journey is not a single gate crossed at launch; it is an accumulation of operational discipline applied through the engineering lifecycle. Teams that treat the production definition as a launch checklist will check the boxes once and move on. Teams that treat it as a living set of acceptance criteria will return to it when the system changes, traffic patterns shift, a new regulatory requirement emerges, or a production incident reveals a gap. The second posture is the one this book is written for.

The Pilot-to-Production Pathway

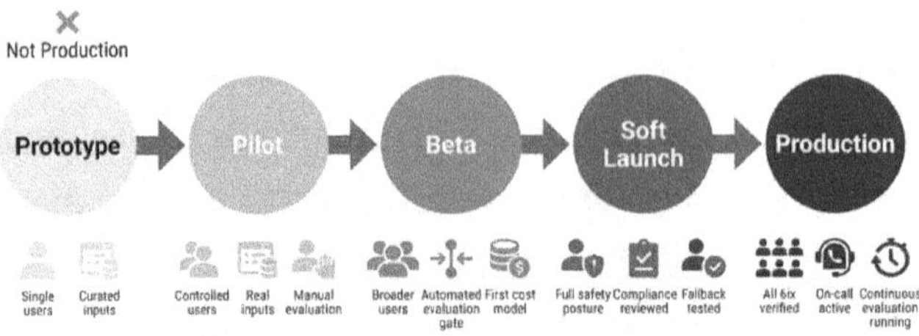

4.13 Technical Checklist

Apply this checklist as a release gate before declaring the LLM feature ready to proceed beyond the scoping phase. Each item corresponds to a decision or artifact that later chapters build on.

1. The team can locate the candidate model on the taxonomy described in this chapter and has documented the reasons the task requires that model category.
2. The alternatives on the taxonomy have been considered and rejected with documented reasons, not simply ignored.
3. The deployment model has been selected using the decision rubric in this chapter, and the constraints driving the selection (latency, volume, data residency, operational maturity) are written down.
4. The alternatives to the selected deployment model have been evaluated, and the reasons for rejection are documented.
5. An ROI estimate exists that names the business value driver (deflection, time saved, automation gain, or revenue), quantifies the value and cost assumptions, and has been reviewed by a finance or business stakeholder.
6. The regulatory regimes the workload falls under have been identified by name, and each has a named owner on the team who is responsible for keeping the assessment current.

7. Internal AI-use policy requirements have been reviewed: vendor approval status, data classification, and any required formal risk assessment.
8. The metric stack is documented with all three layers: quality metrics (correctness, faithfulness, helpfulness, safety), operational metrics (latency, throughput, cost per request, error rate), and business metrics (adoption, deflection, or revenue impact).
9. Each metric in the stack has a named owner and a defined threshold that triggers a review when breached.
10. The four roles (architect, LLM engineer, technical lead, platform partner) are filled by named individuals or teams, and the seams between roles are explicitly documented.
11. The build-buy-compose decision has been made for the current scope, documented with the reasoning, and a review cadence has been set.
12. The six-component production definition has been reviewed, and each component has been assessed as addressed, in progress, or not yet started.
13. The on-call rotation and escalation path for the feature are defined before the first real-traffic deployment.
14. A fallback strategy for primary model endpoint unavailability has been designed and is scheduled for testing before launch.
15. The team can describe what a quality regression would look like for this feature and how it would be detected.

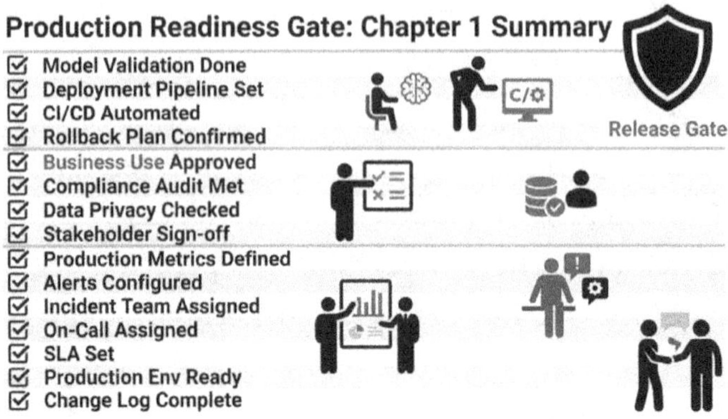

4.14 Team Conversation

Use these questions in a team meeting or a working session to pressure-test the team's alignment on the decisions introduced in this chapter. The questions are designed to surface disagreement and gaps, not to confirm what everyone already believes.

What does 'production' mean for our most important LLM feature, and if two engineers on the team answered that question independently, would their answers match?

Which deployment model are we using, and is that the result of a documented, deliberate decision or of an operational default that was never revisited?

Which of the six LLM-specific failure modes (hallucination, prompt injection, data leakage, model drift, vendor outage, quality regression) have we explicitly planned for, and which are we assuming someone else will catch?

Have we mapped this use case to a regulatory regime, or will we discover the compliance requirements when legal reviews the feature before launch?

Do our success metrics reflect what the business actually values, or what is easiest to instrument from the infrastructure we already have?

Would our finance partners recognize and accept the cost-per-request number we are using in our ROI estimate, and have they reviewed the assumptions behind it?

If our primary model provider had a four-hour outage tomorrow, what exactly would our feature do, and has that fallback path ever been tested under realistic load?

Are the four roles on this project filled by named people, and do those people agree on where their responsibilities end and the next role's responsibilities begin?

When did we last revisit the build-buy-compose decision, and what would need to change in volume, cost, or capability to prompt a revisit?

What is our plan for the first production-quality regression, and who is responsible for detecting, diagnosing, and resolving it?

4.15 Key Takeaway

Large language models have moved past the question of adoption and into the engineering discipline of operations. The teams now building LLM features are not deciding whether to use these models; they are deciding how to operate them with the reliability, cost discipline, safety posture, and observability that the rest of their stack already expects. The landscape established in this chapter supplies the vocabulary and the taxonomy that teams need to make the first set of consequential decisions: which model category, which deployment model, which use-case pattern, and ROI structure, and which metric stack. Each of these decisions has a documented pattern that the team can apply and a clear ownership assignment. None of them requires a research background; all of them require engineering discipline and operational clarity.

The regulatory context and the risk surface are not obstacles to shipping. They are design inputs that, incorporated early, produce a system that is easier to audit, easier to monitor, and cheaper to maintain than a system that discovers compliance and risk requirements after the architecture is fixed. The responsible-by-design principle is not a constraint on the team's ambition; it is the engineering posture that makes that ambition sustainable over the feature's full lifecycle. A system designed with logging, human oversight hooks, and data residency constraints from the start is one that the organization can continue to operate as regulatory requirements evolve, as model capabilities change, and as the traffic profile grows beyond initial projections.

The working definition of 'production' introduced here is the acceptance criterion this entire book tests against. Reliability, observability, safety, cost discipline, evaluation discipline, and a defensible compliance posture are not aspirations; they are engineering deliverables, each with a measurable state and a named owner. The pilot-to-production pathway is the journey from a working prototype to a system that satisfies all six components, and every team on the LLM-in-production path is somewhere on it. This book provides the map for the rest of the journey, chapter by chapter, pattern by pattern, checklist by checklist. The work begins in the next chapter, where the model selection decision is examined with the precision the consequences require.

5 Model Selection and Sizing

5.1 Opening Scenario

The platform lead called the meeting on a Thursday afternoon with a single agenda item: explain the bill. The prior-month invoice listed seven distinct line items from five different model providers, each tied to a different product team. Total spend had grown by 42% in 90 days. The platform team had not sanctioned any of the increases because it had never been consulted on model selection in the first place. Each product team had picked its model independently, usually based on whichever provider had published the most compelling benchmark post in the weeks leading up to that team's kickoff.

The problems surfaced in a pattern. One team had chosen a flagship reasoning model for a task that amounted to structured field extraction; the model's output quality was indistinguishable from a much smaller general-purpose model, but the cost per request was eleven times higher. A second team had selected a provider whose API version they were calling had been quietly deprecated two months after the team shipped; the provider's migration guide required prompt reformatting, which forced a two-sprint rewrite at exactly the moment the team had planned to focus on new features. A third team had self-hosted an open-weight model to reduce API costs, but had not accounted for the GPU memory footprint at the concurrency they needed; they were running at 30% utilization on a configuration that cost twice as much as the managed API would have. The remaining two teams had chosen well, for different reasons, and neither team could fully explain why.

The platform lead's question was simple: how did we choose these models? The answer, repeated five times with minor variation, was that a team member had run a few prompts, the output had looked good, and the model had scored well on a public leaderboard at the time. No team had benchmarked its actual workload. No team had modeled cost at projected volume. No team had documented an upgrade path or asked what would happen when the provider released the next version and retired the current one.

This chapter is the framework that fixes that pattern. It is not a model recommendation; the landscape shifts fast enough that any specific recommendation would be stale before the ink is dry. It is a structured decision process that produces a defensible, documented, cost-aware selection regardless of which models are available when the team is choosing. The platform lead in the scenario above did not need a different answer; she needed a replicable method for arriving at the right answer, whatever that answer turns out to be in any given quarter.

Five-Team Model Selection Fragmentation: From Independent Choices to Platform Incident

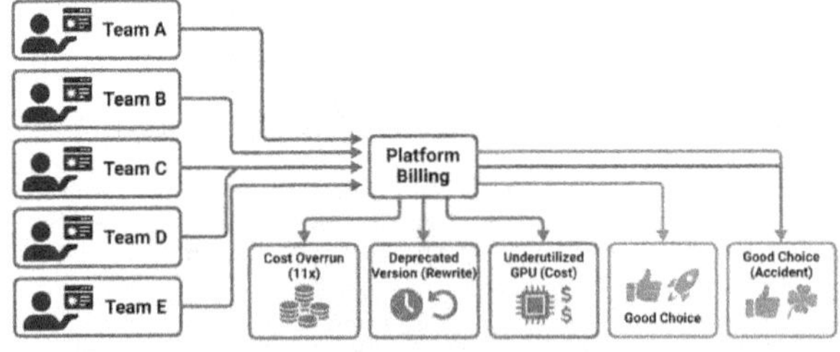

Uncoordinated model selection produces inconsistent outcomes. The two successful teams succeeded by accident, not by process.

5.2 Why It Matters

Model selection is the most upstream decision an LLM project makes, and its consequences compound. The model chosen at kickoff determines the cost curve the team must manage for the next several quarters, the latency envelope available to the product, the fine-tuning strategies that are viable, and the serving architecture the platform team must provision. A mismatch between model capability and task requirements does not announce itself cleanly. It shows up as unexplained quality variance, as gradual cost creep that no single line item explains, and as a brittleness to provider upgrades that is expensive to unwind after the team has built assumptions about model behavior into the product.

The workflow-level impact of a poor selection decision extends well beyond the model itself. Teams that choose without a framework tend to choose again, repeatedly, each time a new provider publishes a compelling benchmark. The result is a codebase layered with prompt formats from three different providers, a serving configuration that has evolved without a coherent sizing rationale, and an on-call rotation that must understand the quirks of whichever model is currently under load. This is not a vendor problem or a market-immaturity problem; it is a decision-process problem, and it has a structural fix.

The mission-aligned goal of this chapter is operational clarity: after reading and applying this framework, every team should be able to answer three questions without hesitation. First, why was this model chosen over its candidates? Second, what cost, latency, and capability of trade-offs did the team accept, and which were measured rather than assumed? Third, what happens when the provider deprecates this version, and has that contingency ever been tested? A team that can answer all three questions has made a responsible selection. A team that cannot is one quarter away from the scenario in the opening.

5.3 Model Families and Their Trade-Offs

The model landscape is large and changes quickly, but the number of meaningful architectural families is smaller and more stable. Rather than tracking individual releases, engineers benefit from understanding the family-level trade-offs, because those trade-offs tend to persist even as individual models within a family are updated. The families that matter for most production decisions cluster into five categories: general-purpose dense models, mixture-of-experts models, small and distilled models, reasoning-specialized models, and domain-adapted or fine-tuned derivatives.

General-purpose dense models are the workhorses of most production deployments. They cover a wide range of capabilities, support long context windows, and are available both as managed API services and as open-weight releases for self-hosting. Their trade-off is straightforward: larger, denser models produce higher-quality output but cost more per token and incur higher latency. Most teams overestimate how large a model they need for their actual workload, which means the first selection pressure on this family should be downward: start with the smallest model that meets quality criteria on representative inputs before justifying a larger one.

Mixture-of-experts models activate only a subset of parameters per forward pass, allowing them to achieve high effective capability with lower per-token compute than a dense model of equivalent total parameter count. The operational implication is non-obvious: the memory footprint of a mixture-of-experts model scales with the total number of parameters, not the number of activated ones, so self-hosting these models requires significantly more GPU memory than their inference cost alone might suggest. Teams choosing between a large dense model and a mixture-of-experts model of similar capability should account for memory as a first-class self-hosting constraint.

Small and distilled models have become a genuinely viable option for a wide range of production tasks. Distillation techniques transfer capability from a large teacher model to a smaller student model, often preserving most of the quality in well-defined task categories at a fraction of the serving cost. For tasks with a narrow and well-defined output space, such as classification, entity extraction, or structured summarization, a distilled model running on commodity hardware frequently outperforms a much larger general-purpose model on the only metrics that matter for production: quality on the actual task, latency at the actual concurrency, and cost at the actual volume.

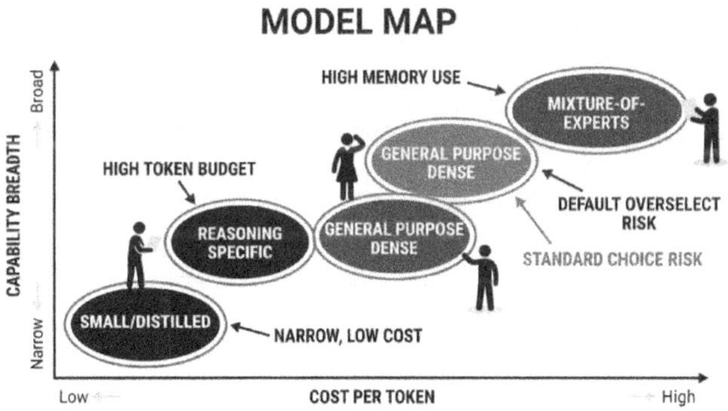

Reasoning-specialized models introduce an entirely different trade-off. These models generate intermediate reasoning steps, sometimes called chain-of-thought tokens or thinking tokens, before producing a final answer. The reasoning tokens are not visible in the final output but are charged as inference tokens by

API providers and consume GPU memory and time during self-hosted inference. The result is that reasoning models can cost an order of magnitude more per request than general-purpose models of similar base size. That cost is justified when the task genuinely requires multi-step logical deduction, formal verification, or complex mathematical reasoning. It is not justified for retrieval, summarization, classification, or conversational tasks where a general-purpose model achieves equivalent output quality at a fraction of the cost.

Domain-adapted or fine-tuned derivatives form the fifth family. These are base models that have been further trained on domain-specific corpora or task-specific labeled datasets. They are relevant when a general-purpose model underperforms on a specialized vocabulary, a specific output format, or a constrained knowledge domain. The selection consideration here is not whether to use a fine-tuned model, but when to fine-tune, prompt engineer, or use retrieval-augmented generation. Chapter 3 covers that distinction in depth; for selection purposes, the key point is that fine-tuned models require their own evaluation discipline and upgrade-path management, since fine-tuning must be reapplied when the base model version changes.

5.4 Open-Source Models versus Managed API Services

The open-versus-managed decision is one of the most consequential choices in the model-selection process and one of the most frequently made on the wrong grounds. Teams that choose self-hosting primarily to reduce cost often find that the operational burden they have taken on, in the form of infrastructure management, version tracking, hardware provisioning, and on-call coverage for the serving layer, exceeds the savings they anticipated. Teams that choose managed APIs primarily for simplicity often find that data-governance constraints, latency requirements, or budget ceilings eventually force a migration that would have been cheaper to plan from the start. The right decision depends on a structured trade-off, not on cost alone.

Managed API services offer a low operational-burden path to production. The provider handles infrastructure provisioning, version management, model updates, safety filters, and rate-limit management. The team pays per token and consumes the capability through a network boundary. The trade-offs are predictable: data leaves the network perimeter on every request, provider rate limits constrain burst capacity, API versioning is at the provider's discretion and not the team's, and unit economics are visible but not controllable below the token price. For teams in their first production deployment, for applications with moderate volume, and for organizations without an existing GPU infrastructure, managed APIs are the default starting point. The pilot-to-production pathway is fastest here.

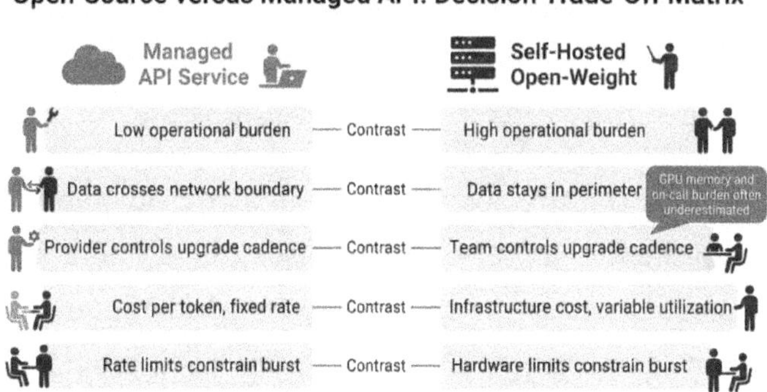

Self-hosted open-weight models shift the operational burden from provider dependency to infrastructure ownership. The team controls the model artifact, the serving configuration, the upgrade schedule, and the data residency. For organizations with strict data-governance requirements, high request volumes where token pricing is prohibitive, or applications that require fine-grained control over inference parameters, self-hosting is the viable path. The hidden costs are well-documented but consistently underestimated: GPU memory requirements scale with model size and concurrent batch sizes; the serving stack introduces its own operational surface, including version compatibility among model weights, the inference engine, and hardware drivers; and on-call coverage must extend to the serving layer itself.

A practical framework for this decision starts with three questions. Does the data in a request have a governance policy that prohibits third-party processing? If yes, the team has no choice; self-hosting is required unless the provider offers a compliant deployment option. At the team's projected request volume, is the per-token cost of the managed API more expensive than the amortized cost of the GPU infrastructure required to self-host the equivalent capability? If not, the cost argument for self-hosting dissolves. Does the team have the platform engineering capacity to operate a model-serving stack with appropriate reliability and observability? If not, the operational burden is a risk that has not been fully priced into the decision.

A pattern that many mature teams adopt is a hybrid: managed APIs for development, lower-volume use cases, and burst capacity; self-hosted models for high-volume steady-state traffic where the economics justify the infrastructure investment. This hybrid pattern requires an abstraction layer between the application and the model endpoint, which is discussed in the section on vendor lock-in later in this chapter. The abstraction layer is not complex for its own sake; it is the mechanism that makes the hybrid pattern operational without forcing every product team to reason about which endpoint is serving their traffic at any given time.

5.5 The Performance, Cost, and Latency Decision Framework

Every model selection decision involves three axes that interact in ways that are not always intuitive. Capability measures how well the model performs on representative inputs. Cost measures the total expense per request at the team's projected volume. Latency measures the time from request submission to usable output, at the percentiles that matter for the product's user experience. These three axes form a triangle: optimizing for any one of them exerts pressure on the other two, and the direction of that pressure is not always the same.

The most common selection error is treating capability performance as the primary criterion and treating cost and latency as constraints to manage later. This produces the pattern seen in the opening scenario: teams select the model that scores best on public benchmarks, then discover during integration that the cost per request is incompatible with their unit economics, or that the model's latency at the concurrency their product requires pushes their P95 response time past the threshold that triggers user abandonment. By that point, the team has built prompt logic, evaluation suites, and downstream processing assumptions around a specific model's output format, and switching carries real engineering cost.

Performance, Cost, and Latency Triangle: Trade-Off Interactions

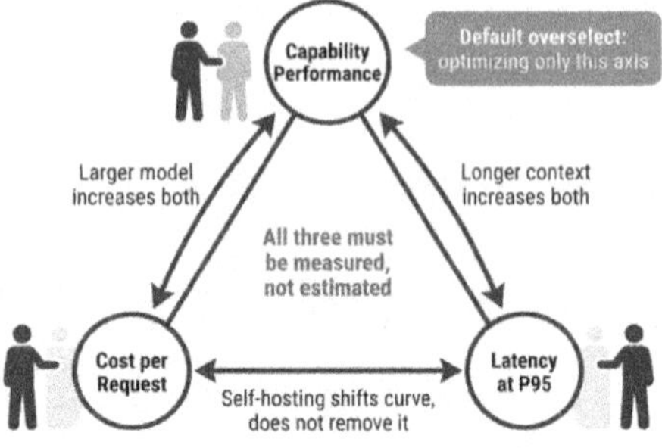

A structured approach to the trade-off triangle begins by establishing the constraints before evaluating candidates. The latency constraint comes from the product: what is the acceptable P95 response time for the user-facing interaction this model will serve? For a synchronous chat interface, the answer is typically under two seconds end-to-end for time-to-first-token, with streaming delivery that masks the remainder. For an asynchronous batch pipeline, the latency budget may be measured in minutes. The constraints differ in kind, and a model well-suited to one is not necessarily well-suited to the other.

The cost constraint comes from the product's unit economics: what is the maximum cost per request that keeps the feature financially viable, given the projected request volume and the revenue or cost reduction the feature is expected to generate? This number should be computed from first principles at selection time, not derived from whatever the chosen model happens to cost. Chapter 6 covers the chargeback and unit-economics framework in depth; for selection purposes, a conservative working estimate is to target a cost per request of no more than 10% of the value the request is expected to generate or save. That number will vary by use case, but having any number is better than having none, because it converts cost from an afterthought into a filter.

With constraints established, the capability criterion becomes a threshold rather than a ranking. The question changes from which model is best to which candidates satisfy the latency and cost constraints, and among those, which achieves acceptable quality on the team's representative workload. This reframe does more than save money; it produces a more durable selection, because models that comfortably exceed quality thresholds on constrained hardware tend to be more robust to prompt variation, edge-case inputs, and the small distribution shifts that occur as real user traffic diverges from the evaluation set over time.

5.6 Capability Benchmarking Methodology

Public benchmark scores are the starting point for model selection, not the ending point. This distinction is important enough to state plainly: a model that tops a public leaderboard on general reasoning tasks may underperform a smaller, cheaper model on the specific task the team needs it to perform. The gap between public-benchmark performance and production-task performance has several structural causes: benchmark tasks often differ in distribution from production inputs; models are sometimes post-trained or prompt-tuned specifically to score well on known evaluation sets; and general-capability benchmarks do not measure the properties that matter most for constrained production tasks, such as structured-output compliance, refusal calibration, or sensitivity to prompt phrasing.

The benchmarking methodology that produces apples-to-apples comparisons for model selection has four components. First, the evaluation set must be drawn from the team's actual workload distribution, not from a public dataset. This means collecting a representative sample of real inputs the model will encounter in production, sanitizing any personally identifiable information if needed, and labeling or curating a ground-truth reference against which model outputs can be scored. The size of this evaluation set should be large enough to produce statistically stable quality metrics but focused enough to be runnable in a reasonable time across multiple candidate models.

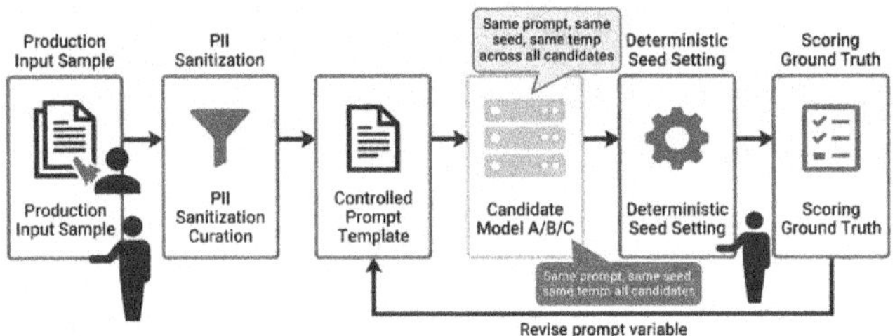

Second, the prompt template used to query all candidate models must be fixed and versioned. A common benchmarking error is tuning the prompt separately for each candidate, leading to a comparison that assesses prompt engineering skill rather than model capability. For a fair comparison, every candidate model receives the same prompt structure, system message, and instruction phrasing. If one model requires a substantially different prompt format to produce acceptable output, that format sensitivity is itself a production-readiness signal, and it should be documented as a capability characteristic rather than accommodated by adjusting the evaluation condition.

Third, inference conditions must be controlled across candidates. When models support temperature and seed parameters, those should be set identically. Where models do not expose a seed, multiple runs should be averaged to account for stochastic variation. The goal is to separate model capability from inference randomness: a model that produces inconsistent output across runs at the same temperature setting is exhibiting a production-relevant property, and that instability should be measured and weighted as part of the selection decision rather than averaged away.

Fourth, the evaluation metric must be specified before looking at model outputs. Choosing the metric after seeing the results introduces selection bias: the team will unconsciously favor metrics that favor the model they have already formed a preference for. Task-specific metrics are preferred over general quality scores: for extraction tasks, precision and recall on the extracted fields; for summarization, factual consistency against a reference; for classification, F1 across all classes; for structured output, schema compliance rate. Human evaluation should be reserved for tasks where no automated metric adequately captures quality, and even then, human evaluation should be calibrated with an inter-annotator agreement check before being used as a selection gate.

5.7 Latency and Throughput Benchmarking

Operational benchmarking is a separate discipline from capability benchmarking and must be run independently. It measures the serving behavior of each candidate model under realistic load conditions, not the quality of its outputs. The metrics that matter are time-to-first-token, which governs the perceived responsiveness of streaming applications; tokens-per-second, which governs throughput efficiency; time-to-last-token, which governs the total latency for non-streaming batch workflows; and the tail-latency profile under concurrent load, which governs reliability under the traffic bursts that production systems routinely experience.

Time-to-first-token is the latency metric most directly felt by users of interactive applications. It measures the elapsed time from the moment the request is submitted to the moment the first token of the response arrives. For a user waiting on a conversational response, a time-to-first-token above approximately 800 milliseconds begins to register as a noticeable pause; above 2 seconds, it noticeably degrades the experience. For applications using streaming delivery, time-to-first-token is the number that must be optimized first, because it determines the perceived responsiveness of the product regardless of how fast subsequent tokens arrive.

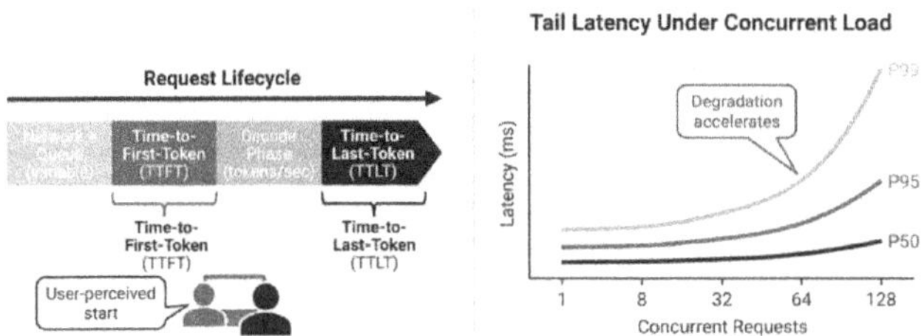

Tokens-per-second is the throughput metric that governs the efficiency and cost-effectiveness of self-hosted infrastructure. It measures how quickly the model generates output tokens after it has started. Higher tokens-per-second means more output per unit of hardware time, which translates directly to throughput per dollar for self-hosted deployments. Managed API providers generally do not expose this metric directly, but it can be inferred by measuring the time between the first and last tokens in a response of known length.

Tail latency under concurrent load is the metric most frequently overlooked during pre-production benchmarking and most frequently responsible for production incidents. Models that perform well at single-request latency often degrade significantly under concurrent load as the serving layer fills its KV cache, queues requests behind long-context completions, or contends for GPU memory. The correct benchmarking approach is to measure latency at the concurrency levels the team expects to sustain in production, not just at a single-request baseline. P95 and P99 latency at the expected peak concurrency should be explicit inputs to the selection decision, not metrics that surface only after the system is deployed.

When running operational benchmarks across candidates, the infrastructure conditions must be controlled as carefully as the prompt conditions in capability benchmarking. For managed API candidates, benchmark runs should be conducted at times representative of expected production traffic, not during off-peak periods when shared infrastructure may be underutilized. For self-hosted candidates, benchmark runs should be conducted on the hardware configuration the team actually plans to deploy, with the batch size and concurrency settings matching the production serving plan. A self-hosted model benchmarked on a larger GPU configuration than the one that will actually serve production traffic will produce latency numbers that are systematically optimistic and cost numbers that are systematically understated.

5.8 Cost Modeling Across Selection Candidates

Cost modeling at selection time requires computing the fully loaded cost per request for each candidate at the team's projected request volume. For managed API candidates, the inputs are straightforward: the provider's token pricing, the expected average input token count per request, the expected average output token count per request, and the projected request volume over a given time period. For self-hosted candidates, the inputs are more complex: the hardware cost per hour, the expected utilization rate, the infrastructure overhead for the serving layer, and the operational cost of the engineering time required to maintain the system.

A common modeling error among managed API candidates is computing cost based on token counts observed in development rather than in production. Development workloads tend to involve shorter prompts, fewer edge cases, and less diverse output lengths than production workloads. A conservative rule is to add a 50% buffer to development-observed token counts when projecting production costs, and to model the scenario where actual volume is twice the projected volume. A system designed to be financially viable at projected volume but not at two times projected volume is not production-ready; it has an undeclared cost ceiling that will eventually be discovered under the best possible conditions, which is to say, when the feature is succeeding and generating more traffic than planned.

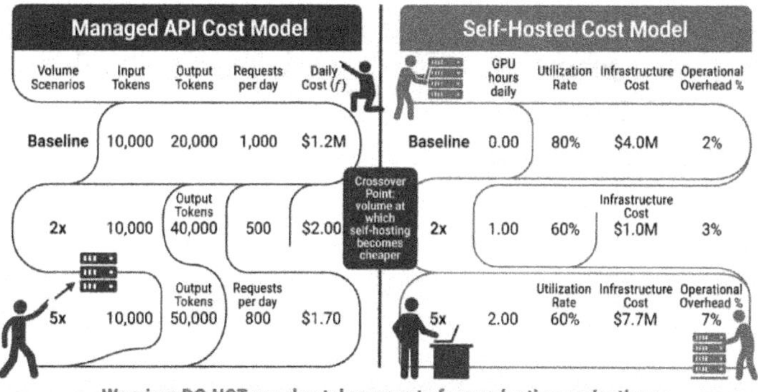

For self-hosted candidates, the utilization rate is the most important and least well-estimated variable. Self-hosted LLM serving infrastructure is rarely utilized at high rates during steady-state operation because traffic is bursty and model inference does not parallelize across requests as efficiently as CPU-bound services. A realistic utilization rate for a production LLM serving deployment ranges from 30 to 60 percent at sustained load; planning for higher utilization without empirical validation is the most common cause of under-provisioning, which produces tail-latency degradation under load. The cost model should be computed at two utilization scenarios: the utilization the team achieves in the first ninety days and the utilization the team believes is achievable with sustained optimization.

The crossover point, where self-hosting becomes cheaper than managed API pricing, is a specific calculation the team should make explicit at selection time. It depends on the provider's token price, hardware cost, and utilization rate, and varies substantially across model sizes and providers. For small and medium workloads under most current pricing regimes, the crossover point is typically in the range of several hundred thousand to several million requests per day, depending on output length.

Teams below that crossover point are almost always better served by managed APIs; teams significantly above it should, at minimum, model self-hosting as an option and explicitly document the infrastructure requirements.

5.9 Evaluation Metrics for Production Readiness

Capability metrics measure what a model can do on a representative evaluation set. Production-readiness metrics measure whether the model is safe and reliable to ship. These are different measurement categories, and passing the first does not imply passing the second. A model can achieve high scores on a capability evaluation while exhibiting output instability across runs, unexpectedly high refusal rates on valid inputs, factual inconsistency between runs, or structured-output non-compliance that breaks downstream parsing. All four of these failure modes have the same effect in production: they make the system unreliable in ways that cannot be fixed by adjusting the model, only by replacing it or adding significant guardrail engineering around it.

Output stability measures the variance in model outputs across multiple runs of the same input at identical inference settings. For deterministic or near-deterministic tasks, high output variance is a signal that the model is sensitive to stochastic effects, which will manifest as inconsistent user experience in production. To measure stability, a representative subset of the evaluation set is passed through each candidate model three to five times with the same temperature and, where available, the same seed. The variance in output quality scores across runs is the stability metric. A model with high capability scores but high output variance requires the team to add consensus or voting logic in the serving layer, which adds latency and complexity that should be priced into the selection decision.

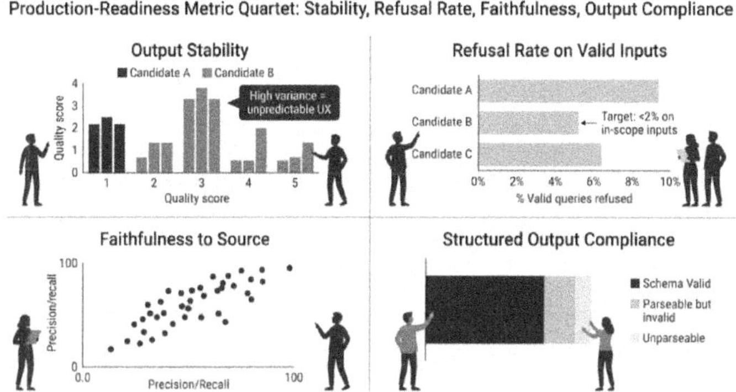

Refusal rate on valid inputs measures how often each candidate model declines to respond to inputs within the application's intended scope. Refusals are expected and correct for genuinely harmful requests, but an elevated refusal rate on valid production inputs is a cost that must be quantified. A refusal means the application must either surface an error to the user, retry with a rephrased prompt adding latency and cost, or fall back to a different model, all of which add complexity. Refusal rate should be measured on the team's actual input distribution, not on a general safety benchmark, because refusal calibration varies significantly by domain and by model post-training approach.

Faithfulness measures how consistently the model's output reflects the content of a provided source when the task involves grounding the response in source material. For tasks such as document summarization, question answering from retrieved context, or data extraction from structured inputs, a model that introduces information not present in the source can hallucinate, producing incorrect outputs in production. Faithfulness is typically scored by comparing each claim in the model output against the source document using either an automated natural-language inference classifier or a careful human annotation pass. It is a distinct metric from overall quality: a model can produce fluent, well-structured output while having low faithfulness to the source, and that combination is worse than lower-quality output that is faithful.

Structured-output compliance measures the rate at which a model's output can be parsed and validated against the schema that the application expects. For applications that use model outputs to populate structured data stores, drive downstream API calls, or render specific UI components, a non-compliant output is an application error that must be caught and handled. The compliance rate should be measured on the evaluation set using the same parsing and validation logic that the application will use in production. A model with 98% compliance on simple schemas may degrade to 80% or 70% compliance on the more complex schemas that production edge cases generate.

5.10 Multimodal and Reasoning Model Selection

Multimodal and reasoning models represent two categories that attract significant selection pressure from product teams, often for the wrong reasons. Multimodal models can process inputs that include images, audio, documents, or structured data alongside text, which makes them genuinely necessary for a class of applications that prior model generations could not support. Reasoning models generate intermediate logical steps that enable them to achieve higher accuracy on tasks requiring multi-step deduction, formal reasoning, or complex problem-solving. Both categories carry higher costs per request than general-purpose text models, meaningfully, and are frequently adopted without a clear justification for those costs.

For multimodal models, the key question is whether the task genuinely requires processing of non-text modalities, or whether the team is considering a multimodal model primarily because its text capabilities are strong. Its multimodal capability comes along for the ride. A multimodal model used exclusively for text tasks is a wasteful choice: the same capability is available in a text-only model at lower cost, and the serving infrastructure for multimodal models is generally more complex and memory-intensive. The selection gate is straightforward: does the production input stream contain non-text content that the model must process as part of its core function? If the answer is no, a multimodal model is not justified.

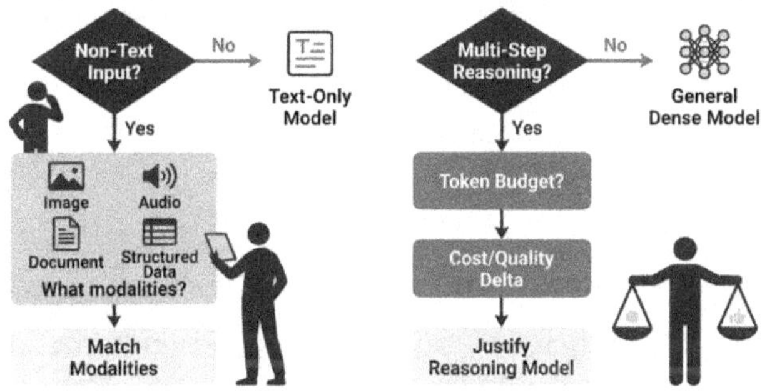

For reasoning models, the justification framework is similar, but the cost signal is more acute. Reasoning models generate extended chains of internal computation that the user does not see, but the team pays for. The reasoning token budget, which is the number of internal tokens the model generates before producing its final answer, varies by task and by model configuration. For complex mathematical or logical problems, the reasoning budget can easily add 500 to several thousand tokens per request, multiplying the effective cost per request and changing the application's unit economics entirely. The correct selection process for reasoning models requires measuring the reasoning token budget on the team's actual inputs, not just on the benchmark tasks that motivated the consideration.

The operational implications of reasoning models extend to latency and cost as well. Because the model generates internal reasoning tokens before producing the final answer, time-to-first-token is extended by the reasoning phase, which can take seconds to tens of seconds for complex problems. For interactive applications where time-to-first-token is a hard user-experience constraint, this extension may be prohibitive regardless of output quality. The selection test is whether the quality improvement achievable with a reasoning model is worth both the cost and the latency increase on the specific task, measured on the specific inputs, at the specific concurrency the production system will operate under.

5.11 Vendor Lock-In, Upgrade Paths, and Multi-Model Strategies

Vendor lock-in in LLM systems takes several forms, some obvious and some subtle. The most obvious form is dependency on a provider's API format: when the provider updates or deprecates an API version, any application that calls that version directly must change its code to migrate. The more subtle forms include dependency on a specific model's output style, which can cause downstream parsing logic to break when a new model version changes its default formatting behavior; dependency on a specific model's capability profile, which can cause evaluation suites and guardrails calibrated to one model to fail silently when a different model is substituted; and dependency on a specific provider's rate-limit and pricing structure, which makes cost modeling for alternatives difficult without a dedicated benchmarking effort.

The standard mitigation for API-level lock-in is an abstraction layer that decouples the application's model invocation logic from any specific provider endpoint. This layer is sometimes called a model gateway or LLM proxy. It accepts requests in a normalized internal format, routes them to the appropriate provider and model based on configuration, and normalizes the response back to the internal format before returning it to the application. The abstraction layer is not a complex component to build; the engineering investment is modest compared to the cost of a forced migration under time pressure. What the abstraction layer cannot protect against is output-style lock-in, which requires additional steps.

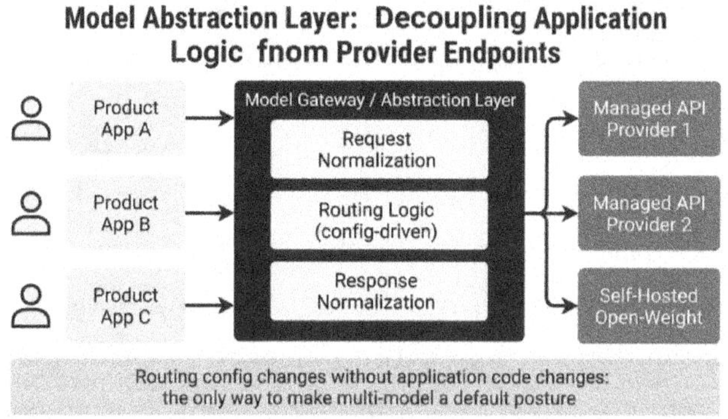

Output-style lock-in requires a model-portability test: a subset of the evaluation set run against the current model and against one or more alternatives, comparing output quality and format compatibility with the downstream processing logic. The portability test should be run periodically, not just at initial selection, because both the model landscape and the downstream logic evolve. Teams that run a portability test annually have a current estimate of the engineering cost to migrate to an alternative model; teams that have never run one face an unknown migration cost, making the actual decision to migrate much harder under pressure.

Upgrade path documentation is a selection criterion in its own right. Before committing to a model version, the team should understand the provider's typical deprecation lifecycle: how much notice they give before retiring a model version, what migration path they provide, and what contingency the team should plan for if the upgrade introduces a quality regression in the team's workload. For managed API providers, this information is usually available in the provider's documentation and release notes. For open-weight models, the upgrade path is entirely the team's responsibility: a new release of the base model requires re-running the capability and operational benchmarks, and if the team has fine-tuned the model, re-running the fine-tuning pipeline as well.

The multi-model strategy is the logical endpoint of the lock-in mitigation work. Rather than selecting a single model as the production model for a given application, a team operating a multi-model strategy maintains two or more serving options and routes traffic between them based on a combination of quality, cost, and availability signals. In its simplest form, this means a primary model and a fallback model: if the primary model is unavailable or returns an error, the serving layer routes the request to the fallback. In more sophisticated configurations, the routing logic is dynamic: requests with specific cost or latency constraints are routed to a cheaper or faster model, while requests that require maximum capability are routed to the flagship model. The abstraction layer described above is the prerequisite for both configurations.

5.12 A Sizing Checklist for Infrastructure and Budget Planning

Sizing translates a model selection decision into a capacity plan: how much infrastructure does the chosen model need to serve the team's traffic at acceptable latency, cost, and reliability? The sizing exercise should be completed before the team declares the selection final, because the infrastructure requirements may affect the selection decision itself. A model that looks cost-effective in isolation may require GPU hardware the organization does not have and cannot procure on the desired timeline, or may require a serving configuration whose operational complexity exceeds the platform team's current bandwidth.

The sizing inputs are the traffic profile, the latency budget, the concurrency assumption, the headroom factor, the growth projection, and the rollback path. The traffic profile specifies the expected request volume by time of day and day of week, the expected distribution of input and output token counts, and the expected peak-to-average ratio. The latency budget specifies the target P50 and P95 time-to-first-token and time-to-last-token for the application tier the model serves. The concurrency assumption specifies the number of simultaneous requests the system must support at peak traffic.

Sizing Input Pyramid: From Traffic Profile to Infrastructure Specification

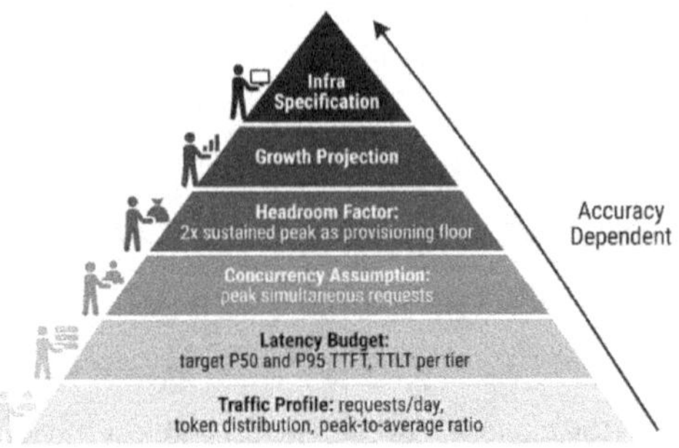

The headroom factor is the provisioning buffer above the expected peak concurrency that the platform team maintains to absorb unexpected traffic spikes without degrading latency. For most LLM-serving deployments, a headroom factor of 2× the expected peak is a reasonable starting point; the exact factor depends on traffic variability and the cost of latency degradation in the specific application context. A customer-facing synchronous application justifies a larger headroom factor than an overnight batch pipeline, because the cost of latency degradation in the synchronous case is user-visible and immediate.

The growth projection is a forward-looking estimate of how request volume will change over the next one to two quarters. Production LLM features almost always see volume growth after launch, driven by expanded user access, new application integrations, and the removal of early beta restrictions. A sizing plan that is exactly right for current volume but has no capacity path to two or three times that volume is not a production-ready plan; it is a plan with a hidden growth ceiling that will manifest as a performance incident when the feature succeeds. The growth projection does not need to be precise; it needs to be explicit, so that the trigger for a capacity review is documented in the sizing plan rather than discovered when the system starts degrading.

The rollback path specifies what the system does if the chosen model must be removed from serving urgently, whether due to a safety incident, a quality regression after a provider update, or an infrastructure failure. For managed API models, the rollback path is typically a pre-configured fallback model in the abstraction layer, which can be activated without a code deployment. For self-hosted models, the rollback path requires either a pinned previous model artifact that can be reloaded without rebuilding the serving configuration or a managed API fallback that provides equivalent capability until the self-hosted serving issue is resolved. The rollback path should be tested at least once before the system goes to production, not planned but untested.

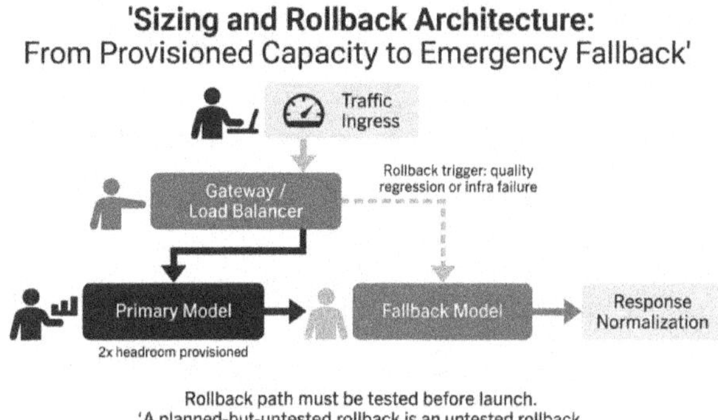

5.13 Technical Checklist

Apply this checklist as a release gate before declaring a model selection production-ready.

1. Two or more candidate models have been benchmarked head-to-head on the team's actual production workload, not on a public evaluation set.
2. The evaluation prompt template is fixed and versioned; every candidate received the same system message and instruction phrasing.

3. Capability, latency, and cost have all been measured on realistic inputs and concurrency levels; none are estimated from marketing claims or single-request baselines.
4. Output stability has been measured across multiple runs at the same inference settings, and high-variance candidates have been downgraded or eliminated.
5. Refusal rate on valid in-scope inputs is below the team's defined threshold for each candidate, measured on the production input distribution.
6. Faithfulness and structured-output compliance have been measured for any task that involves grounded generation or schema-constrained output.
7. The chosen model has a documented upgrade path, including the provider's typical deprecation notice period and the team's migration contingency.
8. Cost modeling at selection time has been computed at baseline volume, two times baseline, and five times baseline, with a fifty-percent token-count buffer applied.
9. The crossover point between managed API and self-hosted cost has been calculated explicitly; the team understands which side of that point it currently sits on.
10. Multimodal or reasoning model capability has been justified rather than defaulted into; the quality-to-cost delta is documented for the specific task.
11. An abstraction layer or model gateway is in place to decouple application code from the specific provider endpoint.
12. A model-portability test has been run against at least one alternative candidate; the estimated migration cost is documented.

13. A sizing document exists for the chosen model, covering traffic profile, latency budget, concurrency assumption, headroom factor, and growth projection.
14. The rollback path has been configured and tested at least once before the system enters production.
15. The sizing document has been reviewed and approved by the platform team.

5.14 Team Conversation

The following questions are designed for use in team meetings to pressure-test alignment on model selection and sizing decisions. They are intended as discussion starters, not as an exhaustive review checklist.

1. How did we choose our current model, and can we reproduce the decision process clearly enough that a new team member could make the same choice today with the same inputs?
2. Which of our current latency and cost numbers were measured on production-representative inputs and concurrency, and which were estimated or carried over from a prototype?
3. What is our deprecation contingency for the current model version, and have we ever run through it, even in a test environment?
4. Are we using a reasoning or multimodal model anywhere in our stack where a smaller or simpler model would achieve equivalent output quality at substantially lower cost?
5. Where in our stack does vendor lock-in present the most risk, and what would an emergency migration cost us in engineering time if we had to execute it under pressure?
6. Do our evaluation benchmarks reflect the distribution of inputs we actually see in production traffic, or do they reflect the distribution we expected when we designed the evaluation?

7. If our sizing assumptions turn out to be off by a factor of two in the direction of higher volume, which component reaches its constraint first, and how long before that happens, would we have visibility into the problem?
8. Have we run a model-portability test against any alternative candidate in the last two quarters, and do we know what the current cost and quality delta between our primary model and its best alternative is?
9. Is our abstraction layer capable of routing to a fallback model without a code deployment, and has that routing been tested end-to-end since the last model version change?
10. What would our cost-per-request look like at five times current volume, and is there a volume threshold at which we would need to reconsider the current selection fundamentally?

5.15 Key Takeaway

Model selection is a structured decision with documented inputs, measured outputs, and a defensible rationale, not a leaderboard pick made during a prototype sprint and never revisited. The teams that avoid the cost overruns, forced migrations, and reliability incidents that attend poor selection decisions are not the teams that happened to choose the right model; they are the teams that applied a repeatable process that produces the right answer regardless of which models are available in any given quarter. The process described in this chapter is a framework: enumerate candidates by family, establish latency and cost constraints before evaluating capabilities, benchmark on your actual workload under controlled conditions, measure production-readiness metrics as a separate gate, model cost across multiple volume scenarios, and size infrastructure before declaring the selection final.

Performance, cost, and latency are inseparable. Optimizing for capability without first establishing cost and latency constraints produces systems that perform well in evaluation but break the budget in production. The selection framework's most important discipline is the refinement of the model from which the best model is selected that satisfies the constraints and, among those, which achieves acceptable quality on the actual task. That reframe is not a concession to frugality; it is the prerequisite for a selection that remains viable as the feature scales and as the provider landscape evolves.

Vendor lock-in awareness is a selection criterion, not an afterthought. The abstraction layer, the portability test, the documented upgrade path, and the tested rollback path are not overengineering for a first deployment; they are the minimum viable lock-in mitigation that converts a dependency into a managed risk. Every team that skips this work is implicitly betting that its chosen model version will never be deprecated, that the provider will never change its pricing or rate-limit structure in a way that requires a migration, and that no quality regression will ever require an emergency swap. That is not a bet that history supports. The sizing checklist and the multi-model strategy are the artifacts that close the loop between selection discipline and operational readiness, and they are the single most effective tools for preventing the scenario in the opening from recurring two quarters later.

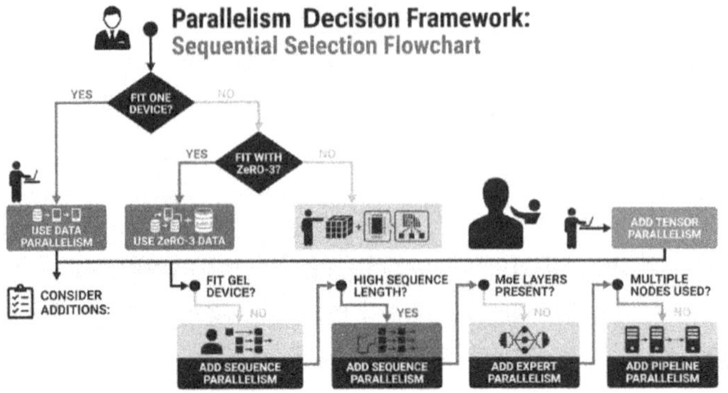
Parallelism Decision Framework:
Sequential Selection Flowchart
FIT ONE DEVICE?
YES
NO
FIT WITH ZeRO-3?
YES
NO
USE DATA PARALLELISM
USE ZeRO-3 DATA
ADD TENSOR PARALLELISM
CONSIDER ADDITIONS:
FIT GEL DEVICE?
NO
ADD SEQUENCE PARALLELISM
HIGH SEQUENCE LENGTH?
YES
ADD SEQUENCE PARALLELISM
MoE LAYERS PRESENT?
NO
ADD EXPERT PARALLELISM
MULTIPLE NODES USED?
NO
ADD PIPELINE PARALLELISM

6 Fine-Tuning and Adaptation

6.1 Opening Scenario

The engineering team at a mid-size software company had been watching the same pattern play out for months: the general-purpose language model they deployed was competent on broad queries but stumbled whenever users asked questions specific to their industry vertical. Response quality dropped sharply in regulatory terminology, internal workflow concepts, and the structured output formats required by downstream systems. The team identified fine-tuning as the solution and allocated six weeks to run it.

The data collection phase moved quickly, perhaps too quickly. An engineer assembled a corpus of internal documents, support tickets, and product manuals, ran a light cleaning pass, and handed the dataset to the training job. The model that emerged from the first successful run genuinely impressed the team in their initial review sessions: it used the correct terminology, matched the expected output schema for test cases drawn from the training set, and, subjectively, felt much sharper than the base model.

Shipment was accelerated. The adapted model went to production after a short internal pilot covering roughly 40 test cases, all drawn from the same corpus used to train it. Within a week, support tickets began accumulating. Users who asked questions outside the precise distribution the model had learned on received confident, fluent, and incorrect answers. General reasoning tasks that the base model had handled reliably now produced degraded output. The team had walked into the two canonical traps of fine-tuning: evaluation coverage that mirrored the training distribution rather than the production distribution, and no held-out general-capability suite to catch regression on everything the model had learned before adaptation began.

The revert took four hours to execute because the serving configuration had been updated in place. Reproducing the partially successful intermediate checkpoint that had shown fewer regressions was impossible: the data preparation steps had not been recorded in a format that could be replayed, the training configuration was embedded in a local script that a developer had modified between runs, and the experiment tracking system showed run metadata but not the dataset version. Six weeks of engineering time produced a model that could not be debugged, reproduced, or safely kept in production.

The failure was not a computational or modeling failure. It was a process failure. No documented decision criteria had established why fine-tuning was the right choice over a retrieval-augmented approach or a more disciplined prompting strategy. No pipeline had been built to make the run reproducible from first principles. No evaluation framework had separated in-distribution test cases from held-out general capability. No rollback plan had been rehearsed. The chapter that follows is the discipline the team should have started with: explicit criteria for when adaptation earns its complexity, the methods that match different problem shapes, the data practices that determine whether a run can be trusted, and the evaluation and deprecation patterns that adapt a repeatable production capability rather than a one-time gamble.

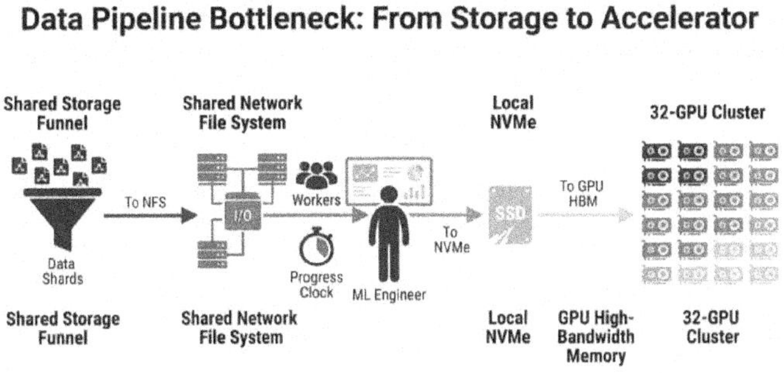

Data Pipeline Bottleneck: From Storage to Accelerator

6.2 Why It Matters

Adaptation sits at the intersection of the two decisions that most directly determine the cost and capability profile of a production LLM system: which model to run, and how to make it behave correctly for the task at hand. The answer to the first question is shaped heavily by the answer to the second. A team that can adapt a smaller base model effectively often operates at a fraction of the inference cost of a team that relies on a larger general model with no task-specific tuning. The workflow-level impact is measurable: reduced token counts, tighter output schemas, faster inference at a given hardware tier, and lower latency for structured extraction tasks where the model has learned the expected format rather than inferring it from a prompt.

The risk profile of fine-tuning is equally consequential. An adaptation that overfits a narrow distribution degrades on everything outside it, often silently: the model continues to return confident, fluent output even as its accuracy on out-of-distribution queries collapses. Detecting this requires evaluation that deliberately reaches outside the training distribution, which most teams do not build until after their first regression incident. The operational cost of an undetected regression in production scales with the fraction of the user surface that touches the adapted behavior and with how long the team takes to notice. Systems that carry no rollback plan pay the full cost.

The adaptation decision also has organizational consequences that extend beyond the initial run. Every fine-tuned model is a versioned artifact that must be maintained, evaluated periodically, and eventually retired. The team that ships an adapter without a deprecation plan inherits an obligation that compounds silently: as the base model version it depends on ages, as the task distribution drifts, and as the data used to train it becomes stale. Adaptation discipline is not only about getting the initial run right; it is about building a system that can be operated, monitored, and eventually replaced without a crisis. Mission-aligned adaptation treats the fine-tuned artifact with the same lifecycle rigor as any other production dependency.

6.3 Full Fine-Tuning versus Parameter-Efficient Methods

Full fine-tuning updates every parameter in the model. On a model with billions of parameters, that means storing and computing gradients for the entire weight matrix on every training step. The compute and memory requirements scale directly with model size, which means that full fine-tuning on models with more than a few billion parameters requires substantial multi-GPU infrastructure, long run times, and significant storage for the resulting checkpoint. The resulting model is a complete, standalone artifact that carries no dependency on the original base model at serving time. That independence is the primary operational advantage of full fine-tuning: the serving path is identical to serving any dense model, with no adapter loading overhead, no adapter compatibility constraints, and no shared base model to version-lock.

The capability shift from full fine-tuning is also the largest available, making it attractive for tasks that require big changes to the model's behavior: new language domains, specialized output formats far from the pretraining distribution, or tasks where the model needs to unlearn broad patterns that conflict with domain-specific ones. But that depth of change is also the source of the primary risk: full fine-tuning is the adaptation method most prone to catastrophic forgetting, the phenomenon in which the model's general capabilities degrade as its weights shift toward the fine-tuning distribution. Reversibility is limited; reverting to base-model behavior requires redeploying the base checkpoint rather than removing an adapter.

Parameter-efficient fine-tuning methods, often grouped under the PEFT umbrella, address the compute, storage, and reversibility constraints by updating only a small fraction of the total parameter count. The base model weights remain frozen. The adapter or modification that encodes the task-specific behavior is a separate, much smaller artifact that is composed with the frozen base at inference time. For most production teams, PEFT is the default approach: the compute requirement is a small fraction of full fine-tuning, the adapter artifact is orders of magnitude smaller than a full checkpoint, and the reversibility story is clean because swapping or removing an adapter does not affect the base model. Full fine-tuning should be treated as a deliberate exception, chosen when the capability shift required genuinely exceeds what PEFT can deliver, and when the team has the infrastructure to run it reproducibly.

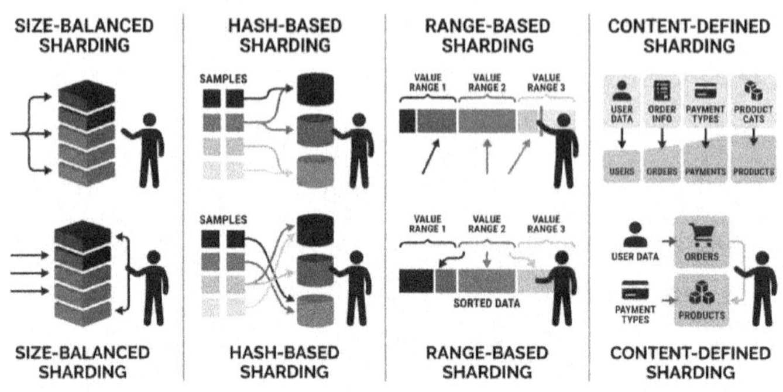

The choice between the two families also interacts with the deployment architecture. Full fine-tuning produces one model per task; serving many fine-tuned variants requires either many replicas of the model or a complex multiplexing layer. PEFT adapters, by contrast, are designed to share a base model across many tasks, which maps naturally to multi-tenant serving architectures and per-customer customization patterns. That serving-time flexibility is a significant operational advantage when the number of distinct task variants grows beyond a handful.

6.4 LoRA, Adapters, Prefix Tuning, and Modern PEFT Families

Low-rank adaptation, widely referred to as LoRA, is the most operationally prevalent PEFT method in production systems as of this writing. The core mechanism is straightforward: instead of updating the full weight matrix W, LoRA learns two low-rank matrices A and B whose product approximates the update to W. The rank parameter, typically written as r, controls the size of this approximation. At r equal to 4 or 8, the adapter for a multi-billion-parameter model may contain only a few million trainable parameters, making training fast, memory-efficient, and storage-cheap. At inference time, the product of A and B can be merged directly into the base weight matrix, eliminating any runtime overhead. That merge capability makes LoRA-adapted models usable with any serving stack that can load the base model, without requiring special adapter-aware inference code.

QLoRA extends LoRA to operate on a quantized base model. The base weights are held in a 4-bit or 8-bit quantized representation, which dramatically reduces the GPU memory required to hold the frozen base during training. LoRA adapters are trained at higher precision on top of the quantized base. The result is a fine-tuning method that makes large-model adaptation accessible on hardware tiers that could not otherwise fit the model in memory. The trade-off is a modest accuracy gap relative to full-precision LoRA, which must be measured against the specific task requirements before committing to the quantized training path.

Prefix tuning and prompt tuning occupy different points in the PEFT design space. Rather than modifying model weights, these methods prepend a learned sequence of continuous vectors to the model's input or to the key-value activations at each layer. The model weights remain completely frozen; only the prefix parameters are trained. The practical effect is that the model's attention mechanism is steered by the learned prefix at every layer, shaping its behavior toward the target task without any weight update. The adapter is extremely small (a few thousand to a few hundred thousand parameters), and multiple prefixes can be hot-swapped at inference time by simply substituting the prefix vectors. The limitation is the capability ceiling: prefix methods tend to underperform LoRA on tasks that require significant behavioral change, particularly those far from the base model's pretraining distribution.

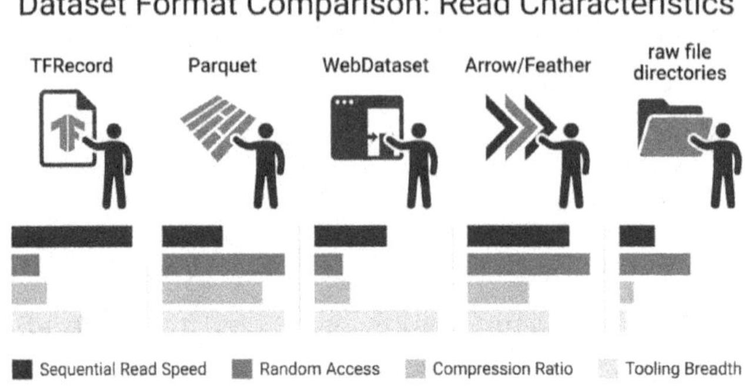

Adapter layers, the earliest PEFT family to see broad adoption, insert small bottleneck modules between the transformer's existing layers. Each adapter module projects down to a lower-dimensional space, applies a nonlinearity, and projects back up. The base model layers remain frozen; only the adapter modules are trained. Adapter layers can achieve a higher capability shift than prefix methods because they modify the activation flow at every layer rather than only prepending steering vectors. Still, they add latency at inference time when not merged, because the adapter forward pass cannot be eliminated as easily as with LoRA. For latency-sensitive serving paths, unmerged adapters require careful benchmarking.

Modern variants continue to refine these families. DoRA decomposes weight updates into magnitude and direction components and trains them separately, recovering some of the expressiveness of full fine-tuning at PEFT parameter counts. IA3 learns scaling vectors for attention and feedforward activations, achieving even smaller adapters than LoRA, albeit at the cost of a lower capability ceiling. The operational guidance for selecting among these variants is to benchmark the specific task rather than to rely on general leaderboard rankings, because relative performance across PEFT methods varies substantially with task type, base model architecture, and dataset size.

6.5 Continued Pretraining versus Instruction Tuning versus Preference Tuning

Adaptation methods differ not only in which parameters they update but in what signal they train on and what behavior they are trying to shift. Three distinct regimes serve three distinct problem shapes, and conflating them is a common source of wasted runs and confusing results.

Continued pretraining trains the model on raw text using the same next-token prediction objective as the original pretraining run. The goal is to shift the model's knowledge distribution toward a domain that was underrepresented in the original pretraining corpus: a specialized scientific field, a proprietary technical vocabulary, a low-resource language, or a large body of internal documentation. Continued pretraining does not change how the model responds to instructions; it changes what the model knows. A model that has undergone continued pretraining on a medical corpus will use clinical terminology more accurately and hallucinate less on clinical facts. Still, its instruction-following style will remain essentially unchanged from the base model. This regime is appropriate when the primary gap between the base model and the production requirement is factual coverage of a specialized domain, not behavioral style.

Instruction tuning trains the model on input-output pairs formatted as instructions and responses. The training signal teaches the model to follow the structure, tone, and constraints of the instruction format that the production system will use. A model that has undergone instruction tuning learns to produce structured JSON when asked for it, to follow length constraints when specified, to adopt a particular persona consistently, or to decline categories of requests predictably. The quality of instruction tuning is almost entirely determined by the quality of the instruction-response pairs in the training set: noisy, inconsistent, or mislabeled pairs produce a model that is noisy, inconsistent, or misaligned in production. This is the regime that most teams who describe 'fine-tuning on internal data' are actually running, and it is the most sensitive to data-curation quality.

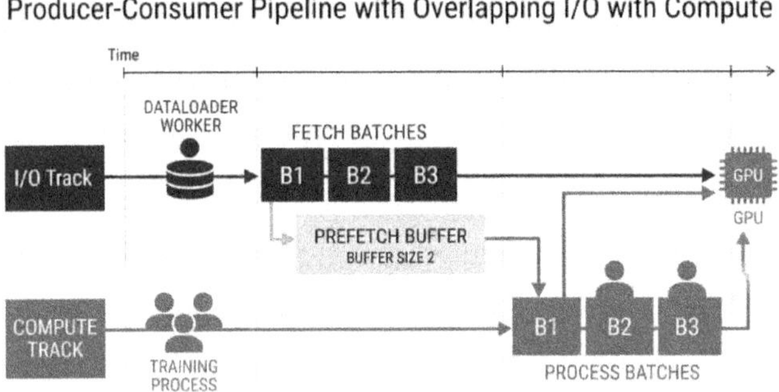

Preference tuning, commonly implemented via direct preference optimization or reinforcement learning from human feedback, trains the model to prefer outputs that human judges or a reward model score higher over outputs they score lower. Rather than specifying correct outputs directly, preference tuning provides the model with ranked pairs: one preferred response and one rejected response for the same prompt. The model learns to shift its output distribution toward the preferred responses. This regime is appropriate when the gap between the model's current behavior and the desired behavior is a question of judgment rather than fact: response tone, refusal behavior, helpfulness under ambiguity, and alignment with organizational values that are difficult to specify as explicit rules. Preference tuning requires more expensive data collection than instruction tuning because ranked pairs require human comparison judgments rather than single correct answers, and the reward signal is noisier.

The three regimes are not mutually exclusive and are often applied in sequence. A common production pipeline runs continued pretraining to establish domain knowledge, instruction tuning to shape task behavior, and then a preference tuning pass to align the model's judgment with organizational standards. Each successive stage benefits from treating the previous stage's output as its starting checkpoint rather than the original base model, and each stage requires its own evaluation suite to verify that it achieved its intended effect without regressing on the gains from prior stages.

6.6 Data Curation, Labeling, and Augmentation Strategies

Data quality is the dominant lever in adaptation outcomes. A well-chosen PEFT method with high-quality training data will outperform a well-tuned full fine-tuning run on poor-quality data. The converse is also true: even full fine-tuning cannot reliably recover from a noisy, duplicated, contaminated, or inconsistently labeled dataset. Most teams discover this the hard way, after a run that costs significant compute produces a model whose behavior cannot be explained by the training configuration because the real explanation is buried in the data.

Deduplication is the first and most important step in cleaning. Training data for instruction tuning frequently contains near-duplicate examples drawn from the same source documents: the same question phrased slightly differently, the same template filled with different values, or the same document chunked in overlapping windows. Near-duplicate training examples bias the model toward the patterns they share, inflate the effective dataset size in ways that do not generalize, and can create memorization artifacts that surface as verbatim recall of training text in production. Exact deduplication is a minimum requirement. Near-deduplication using MinHash or embedding-based similarity is the standard in production.

Tiered Storage Architecture

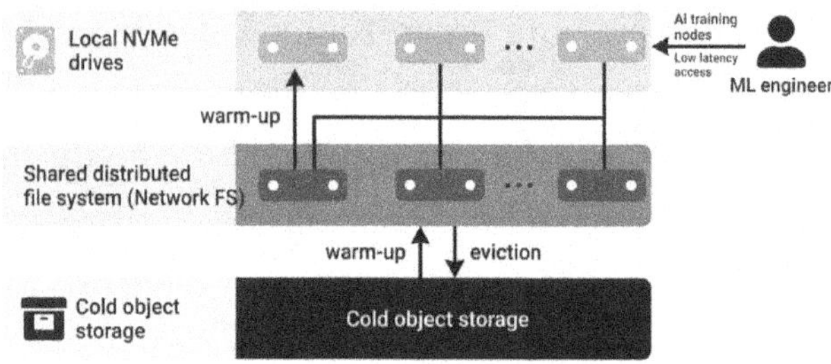

Decontamination requires removing examples that overlap with the evaluation sets the team plans to use to validate the model. A training dataset that contains examples semantically similar to held-out evaluation questions will produce inflated metrics that do not reflect real-world capability. Decontamination against standard benchmarks (general capability suites, public instruction-following benchmarks) is achievable by comparing against known benchmark prompts. Decontamination against internal evaluation sets requires the evaluation set to be finalized before data collection begins, which is a process constraint that most teams violate by constructing both in parallel.

Quality filtering applies heuristic and model-based signals to remove low-quality examples: responses that are too short to be informative, responses that contain formatting artifacts from web scraping, responses that are factually inconsistent with the domain the model is adapted to, or instruction-response pairs in which the response does not actually address the instruction. Heuristic filters are fast and cheap. Model-based quality classifiers are more accurate but require a bootstrapping step. A common pattern is to run heuristic filters first, then use a separately trained quality classifier on the filtered dataset to perform a second pass that catches the examples heuristics miss.

Balanced sampling ensures that the training distribution does not overrepresent a narrow slice of the target task. A dataset built by exporting all support tickets from the past three years will contain far more examples of common request types than rare but important ones. The model trained on this dataset will perform well on common cases and poorly on the long tail, which is often exactly where the production risk is highest. Balancing by category, by source, and by query complexity requires knowing the target distribution well enough to identify which categories matter, which is itself a product of deliberate data analysis rather than automatic pipeline construction.

Labeling quality for instruction tuning determines the ceiling on what the model can learn. Inconsistent labels, where the same prompt receives different responses across training examples due to different annotators or annotation guidelines, introduce noise that the model cannot resolve. The model learns a blurred average of the inconsistent signal rather than any coherent behavior. Annotation guidelines must be specific enough to produce consistent responses from different annotators on the same prompt, which means they must address not just the general task but also the edge cases, the format requirements, and the handling of ambiguous inputs. Inter-annotator agreement measurement is the operational signal for guideline quality.

6.7 Building Reproducible Fine-Tuning Pipelines

Reproducibility is the property that allows a team to re-run a fine-tuning experiment and arrive at the same result, to share a run with a colleague who joins later, to audit a past run when a production issue surfaces, and to iterate on a successful checkpoint without losing track of what made it successful. Most teams discover they lack reproducibility only after they need it: when a regression appears in production and the training artifacts needed to diagnose it are no longer usable.

Versioned datasets are the foundation. The training data must be stored as an immutable artifact with a content-addressable identifier, not as a reference to a directory that can change between runs. The identifier must appear in the run manifest so that any future run referencing the same dataset identifier is guaranteed to train on the same data. This requires a dataset registry that stores dataset versions as first-class objects rather than as filesystem paths on a shared drive. Teams that skip this step frequently find that their 'same dataset' runs are actually training on slightly different data because the source directory was modified between runs.

Checkpointable Streaming Pipeline for Foundation-Model Training

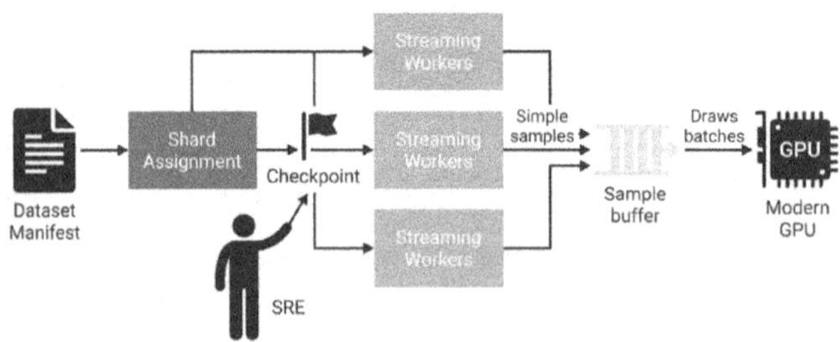

Deterministic data ordering controls a subtle source of variability. Shuffling training data with a fixed random seed produces a consistent ordering across runs with the same seed. Shuffling with a wall-clock time seed or a system entropy source produces a different ordering in each run, which means two runs with identical data and configuration can produce slightly different models due to ordering effects on the stochastic gradient dynamics. The seed must be recorded in the run manifest and set explicitly before the data loader initializes.

Immutable run manifests capture the complete set of inputs needed to reproduce a run: the dataset version identifier, the base model version identifier, the training configuration (learning rate, batch size, number of steps, optimizer parameters, adapter rank, and target layers), the random seeds, and the software environment version. A run manifest is not a training script; it is a snapshot of the values that the training script consumed. Scripts can be refactored; manifests must be preserved. The manifest should be stored as a structured artifact alongside the resulting model checkpoint, not in a separate experiment-tracking system that might lose the link.

Experiment tracking systems provide the search and comparison layer on top of individual run manifests. They allow a team to compare loss curves across runs, filter runs by hyperparameter value, and identify which configuration produced the best checkpoint for a given metric. The operational requirement is that the experiment tracker must store the run manifest, not just the loss curve: a run record that contains metrics but not the dataset version or configuration that produced them is useful for visualization, but cannot support reproduction. Personnel changes are the most common failure mode: the engineer who remembers which run used the cleaned dataset departs, and experiment-tracking entries without embedded manifests become uninterpretable.

6.8 Evaluation and Validation for Adapted Models

Evaluation for adapted models requires two distinct test sets that serve different diagnostic functions. Collapsing them into one is the most common evaluation mistake in adaptation work and the direct cause of the production regression pattern described in the opening scenario.

The in-distribution test set verifies that the adaptation succeeded at its stated objective. It contains examples drawn from the same distribution as the training data, held out from training by a strict split applied before data collection begins. Performance on the in-distribution test set answers the question: Did the model learn the target task? Metrics on this set should be reported alongside a baseline from the un-adapted base model and a ceiling estimate from the best available reference. If the adapted model does not outperform the base model on the in-distribution test set, the adaptation has not achieved its goal and should not proceed to production.

Deduplication and Shuffle Pipeline at Scale

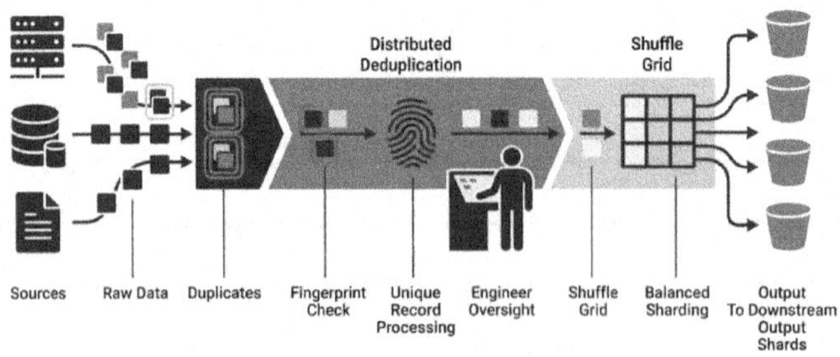

The general-capability held-out suite addresses the orthogonal question: did the adaptation break something the model was already good at? This set contains tasks that are deliberately outside the fine-tuning distribution: general reasoning, instruction following on novel prompt formats, factual recall on topics unrelated to the fine-tuning domain, and common sense inference. The goal is not to measure task-specific improvement; it is to detect regression. A model that scores well on the in-distribution test set but regresses by five percentage points on general reasoning is exhibiting catastrophic forgetting. That regression will not appear in any evaluation that covers only the fine-tuning domain.

Side-by-side comparison with the base model is the most interpretable evaluation format for stakeholders who need to understand the trade-off before approving a production deployment. The comparison presents the base model output and the adapted model output for the same prompt, labeled only as Model A and Model B, and asks human evaluators to rate preference on the target task dimensions. Win rate on the in-distribution task combined with regression rate on the general-capability suite gives the team a two-dimensional view of the adaptation trade-off that neither set alone can provide.

Calibration evaluation is a frequently neglected dimension specific to adapted models. Fine-tuning on a narrow distribution can shift the model's confidence away from its actual accuracy: a model that was well-calibrated on the base distribution may become overconfident on in-distribution queries and underconfident on out-of-distribution ones after fine-tuning. Expected calibration error measured over the general-capability suite provides an early signal for this pattern before it manifests as confident hallucination in production.

6.9 Synthetic Data and Distillation Patterns

Synthetic data generation has become a standard component of the adaptation data pipeline for most production teams. The core pattern uses a larger or more capable model to generate labeled examples for training a smaller, more targeted model. The larger model acts as a labeling engine. Given a prompt template and a set of seed inputs, it generates instruction-response pairs at scale, producing thousands of training examples that would have required human annotators to produce at significantly higher cost and latency. The resulting training data often better capture the target task distribution than organic data collection, because the seed inputs can be designed to cover the long tail of input variations the model will encounter in production.

Self-instruct workflows extend this pattern by using the model being fine-tuned as its own labeling engine in a bootstrapping loop. The model is prompted to generate new instruction examples given a small set of seed instructions, and the generated instructions are filtered for quality and diversity before being added to the training set. The bootstrapping loop repeats until the training set reaches the target size. Self-instruct can produce large, diverse instruction datasets from a small initial seed. Still, it inherits all of the biases and errors of the model it uses as the generator, and quality filtering must be applied aggressively to prevent the model from training on its own worst outputs.

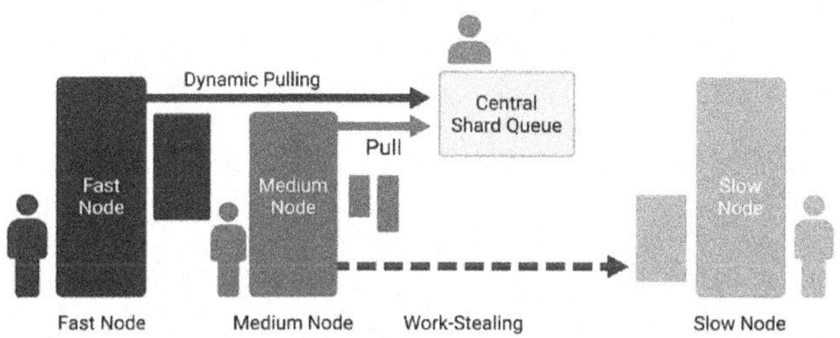

Distillation is a related but distinct pattern in which a student model is trained not on hard labels (correct answers) but on the soft probability distributions (logits) produced by a larger teacher model. The student learns to match the teacher's output distribution, which encodes more information than a hard label: the teacher's confidence, its ranking of alternative outputs, and the similarity structure of its prediction space. Knowledge distillation can produce student models that outperform models of the same size trained solely on hard labels. Still, it requires access to the teacher model's output logits during training, which is not always available when the teacher is an API-only model.

The contamination risk in synthetic data pipelines requires careful management. A large model used as a labeling engine has been trained on public benchmarks. If the seed inputs used to drive synthetic generation overlap with evaluation benchmarks, the generated examples may encode benchmark-specific patterns that inflate evaluation scores on those benchmarks without reflecting real production capability. Decontaminating synthetic data against the evaluation suite is therefore as important as decontaminating organic data, and it requires the same discipline: the evaluation suite must be finalized before synthetic generation begins.

6.10 Catastrophic Forgetting and Mitigation Patterns

Catastrophic forgetting is the dominant failure mode of fine-tuning and the one most likely to reach production undetected. The mechanism is straightforward: as gradient updates shift the model's weights toward the fine-tuning distribution, they move away from the configurations that encoded the model's prior capabilities. The model forgets how to do things it could do before. The failure is silent in the absence of a general-capability evaluation because the model's performance on the fine-tuning task continues to improve even as it regresses on everything else. A narrow evaluation suite captures only fine-tuning task performance and reports improvement; regression in general capabilities remains invisible.

The severity of forgetting scales with the degree of distribution shift between the fine-tuning data and the pretraining data, with the learning rate, and with the number of fine-tuning steps. High learning rates applied over many steps on a narrow dataset lead to the most severe forgetting. Low learning rates, shorter training runs, and PEFT methods that freeze the base weights all help mitigate the problem. PEFT is the most reliable structural mitigation: an adapter that adds task-specific behavior on top of frozen weights cannot, by construction, overwrite the base model's encoded capabilities. The catastrophic forgetting risk in PEFT is therefore substantially lower than in full fine-tuning, though not zero: adapter interference in the residual stream can still degrade base-model behavior to some extent.

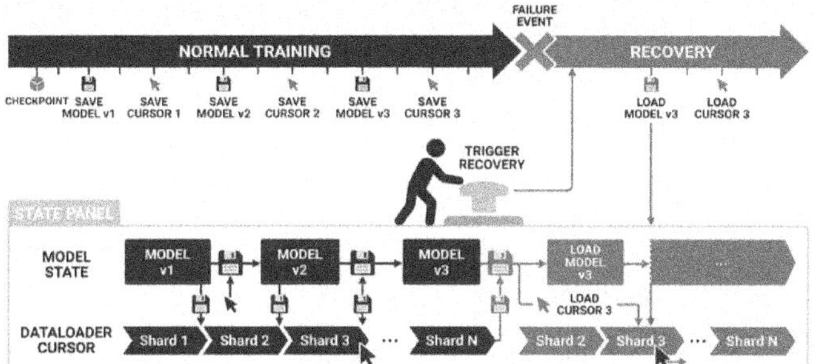

Rehearsal is the mitigation pattern most applicable to full fine-tuning scenarios where PEFT is not an option. The rehearsal approach adds a small percentage of examples drawn from the original pretraining distribution (or a representative proxy) into the fine-tuning batch. The model continues to receive gradient signal from the original distribution even as it trains on the new one, which prevents the weights from drifting too far. The fraction of rehearsal examples is a hyperparameter that must be tuned: too low, and it fails to prevent forgetting; too high, and it dilutes the fine-tuning signal and slows convergence.

Elastic weight consolidation provides a regularization-based alternative. The method estimates the importance of each weight to the original task (using the Fisher information matrix or an approximation thereof). It adds a penalty term to the fine-tuning loss that discourages large changes to important weights. Weights that were critical to the original task's performance are protected from large updates; less important weights can shift more freely. The computational overhead of estimating weight importance is non-trivial and scales with model size, making this approach more practical for smaller models. For production teams operating at large model scale, the combination of PEFT with a well-calibrated learning rate is usually the more practical path.

6.11 Multi-Adapter Serving and Per-Tenant Customization

The economics of adapter serving change materially as the number of distinct task variants grows. A team managing two or three fine-tuned models can operate separate model replicas for each variant without significant overhead. A team managing forty task variants, or a platform serving hundreds of customers each with their own customization, cannot afford to run forty or hundreds of model replicas. The multi-adapter serving pattern resolves this by holding a single base model in GPU memory and loading adapters on a per-request or per-tenant basis, dramatically reducing the memory footprint per served variant.

Dynamic adapter loading requires a serving infrastructure that understands the adapter format and can swap adapters between requests without reloading the base model weights. The serving system maintains a base model resident in GPU memory and an adapter cache in host memory or on fast storage. When a request arrives for a specific adapter, the system loads the adapter weights into GPU memory, applies them to the appropriate layers, processes the request, and either caches the adapter for subsequent requests from the same tenant or evicts it to make room for the next adapter. Cache hit rate is the primary operational metric: a system where adapter loading dominates request latency because cache hit rates are low has a routing or batching problem, not a capacity problem.

Data Quality Gate in the Training I/O Path

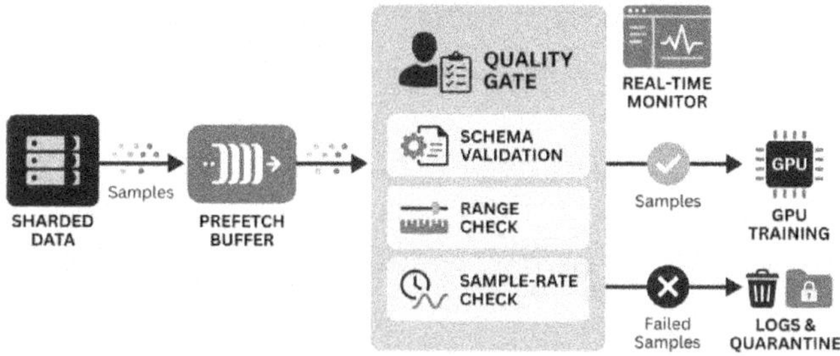

Per-tenant adapter routing introduces isolation requirements. A request from Tenant A must not receive output influenced by Tenant B's adapter, even when both adapters are resident in the serving system simultaneously. For LoRA adapters, this isolation is enforced at the mathematical level: the adapter weights are tenant-specific and applied only to the designated request. The risk of cross-tenant contamination exists at the infrastructure level rather than the model level: shared context windows, shared KV cache entries, or mis-routed batches can cause adapter leakage without any error in the adapter mathematics. Audit logging of which adapter was applied to which request is an operational requirement for multi-tenant deployments.

Adapter composition, the pattern of applying multiple adapters simultaneously, enables more flexible customization hierarchies. A platform might maintain a base model, a domain-level adapter shared across all tenants in a vertical, and a tenant-specific adapter applied on top of the domain adapter. The composition of LoRA adapters is mathematically well-defined for the non-overlapping case and requires careful handling when adapters target the same weight matrices. The operational implication is that composed adapters must be tested together, not just individually: an adapter pair that performs well in isolation can produce unexpected behavior when combined due to interactions at shared weight matrices.

6.12 When to Fine-Tune versus When to Prompt versus When to Retrieve

The decision among fine-tuning, prompting, and retrieval-augmented generation is not a once-and-for-all architectural choice; it is a recurring decision that should be revisited as the system matures, as the task distribution shifts, and as the cost of each option changes with improvements in model capabilities. The discipline is to apply explicit criteria at each decision point rather than defaulting to fine-tuning because it feels more thorough, or to prompting because it feels simpler.

Prompting alone is the right starting point for any new task. The cost of engineering a high-quality prompt is low relative to that of a fine-tuning run, and modern base models respond surprisingly well to carefully crafted few-shot examples and explicit constraint specifications. The team should exhaust the prompting option before committing to fine-tuning. The signal that prompting has been exhausted is not 'we tried a few prompts, and they did not work'; it is 'we have applied systematic prompt optimization, we have measured performance on a held-out test set, and we have hit a ceiling that the task requirements require us to exceed.'

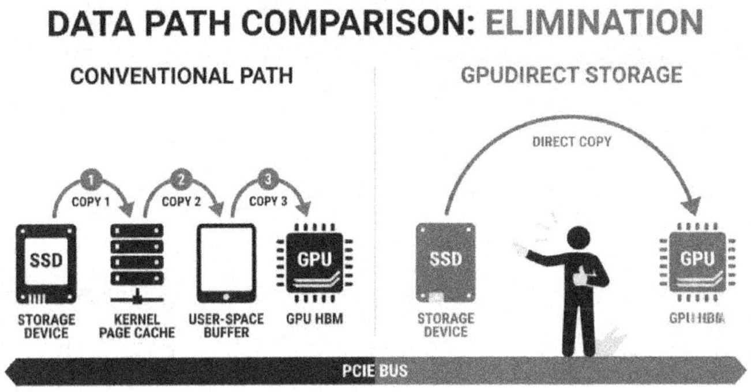

Retrieval-augmented generation is the right choice when the gap between the model's performance and the task requirement is primarily a knowledge gap involving dynamic information: facts that change frequently, documents that are added after the model's training cutoff, proprietary information that cannot be included in training data due to access or privacy constraints, and information that is too voluminous to encode efficiently into model weights. Retrieval does not improve the model's reasoning capability or its behavioral style; it extends its effective knowledge surface without requiring a training run. The combination of retrieval with fine-tuning is appropriate when the task requires both extended knowledge access and a behavioral style shift that prompting cannot achieve.

Fine-tuning earns its complexity when at least one of three conditions holds: the task requires behavioral patterns that are far from the base model's pretraining distribution and cannot be reliably elicited by prompt engineering; the inference cost of a large base model with a long system prompt is too high for the production budget and a smaller adapted model can achieve equivalent task performance; or the task requires output format compliance at a level of consistency that prompt-based instructions cannot reliably enforce at production query volume.

If none of these conditions holds, the complexity of fine-tuning adds risk and operational overhead without proportional benefit.

The decision framework should be documented at the time the decision is made, not reconstructed later. The documentation should record which conditions were evaluated, which data supported the evaluation, and who approved the decision. This record becomes essential when the team revisits the decision six months later, after the model capability landscape has shifted, and what required fine-tuning in the previous cycle can now be achieved by prompting a more capable base model.

6.13 A Rollback and Deprecation Plan for Adapted Models

Adapted models accumulate silently. A team that ships one adapter per quarter has four adapters in production after one year, eight after two. Without an explicit deprecation process, that accumulation grows without bound: old adapters continue to serve traffic because no one has a reason to turn them off, even as the base model they depend on ages, the task distribution they were trained on drifts, and the data that produced them becomes stale. The operational cost of legacy adapters is not just the serving resources they consume; it is the maintenance obligation they create whenever the base model version needs to be updated or when a security patch requires retraining.

A rollback plan should be written before the adapted model ships, not improvised when a regression appears in production. The plan must specify: the version of the base model or prior adapter that will be reinstated if the new adapter must be reverted; the procedure for switching serving traffic back to that version, including any configuration changes required; the target time-to-rollback, measured from the moment the decision to revert is made to the moment the prior version is receiving one hundred percent of production traffic; and the name of the person authorized to execute the rollback without additional approval during an incident. A rollback plan that requires a committee decision during an incident is not a plan.

I/O Optimization Decision Tree

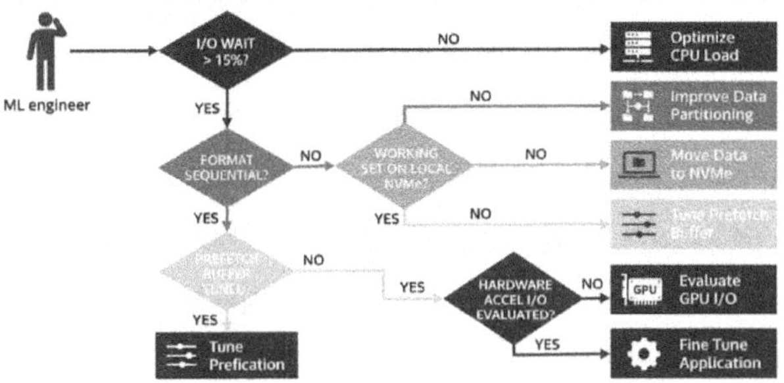

Scheduled re-evaluation is the operational pattern that prevents adapted models from quietly outliving their usefulness. Each adapter should have a re-evaluation schedule set at ship time: typically quarterly for adapters serving high-traffic paths and semi-annually for lower-traffic ones. The re-evaluation runs the adapter against the current version of both the in-distribution test set and the general-capability suite, comparing results to the baseline established at ship time. An adapter that no longer outperforms the current base model on the in-distribution task (because the base model has improved) should be flagged for replacement or retirement rather than maintained indefinitely.

Deprecation notices give consuming teams time to adapt. When a scheduled re-evaluation determines that an adapter will be retired, the teams or systems that call that adapter endpoint should receive advance notice with enough lead time to test against the replacement. A deprecation process that removes an adapter without notice creates the same class of incident as removing any other production dependency without notice. The adapter registry should track which systems depend on which adapter versions, so that the deprecation notice can be routed to the right owners automatically rather than requiring a manual audit of which teams are affected.

Adapter retirement is an archival operation, not a deletion operation. The adapter artifact, the run manifest that produced it, the evaluation results at ship time, and the deprecation timeline should all be preserved in cold storage after the adapter is removed from active serving. Incident investigations, compliance audits, and future training runs may all require access to the historical record. A deletion policy that removes training artifacts after a fixed retention window should be applied only after the compliance obligations for that artifact have been verified and documented.

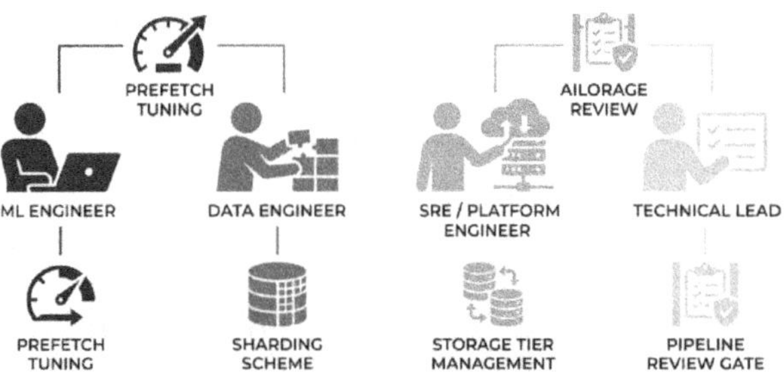

6.14 Technical Checklist

Apply this checklist as a release gate before declaring a fine-tuned or adapted model production-ready. Each item represents a condition that, if unmet, introduces a known class of risk into the deployed system.

1. The decision to fine-tune rather than prompt or retrieve is documented with explicit criteria evaluated against the current task requirements and model capability baseline.
2. The chosen PEFT method is matched to the problem shape: LoRA or QLoRA for most behavioral adaptation, prefix tuning only when the capability shift required is modest, full fine-tuning only when PEFT cannot reach the required capability shift.
3. The training dataset is versioned with a content-addressable identifier recorded in the run manifest. The same identifier will produce the same dataset on any future replay.
4. The dataset has been deduplicated using at least exact deduplication, with near-deduplication applied for datasets assembled from web or document sources.
5. The evaluation suite is split into an in-distribution test set and a general-capability held-out suite. Both sets were finalized before data collection and training began.
6. Decontamination has been performed: the training data has been checked against both the in-distribution test set and the general-capability suite for semantic overlap.
7. The adapted model has been compared to the base model on both test sets. The general-capability suite shows no regression beyond the agreed threshold.
8. Calibration has been measured on the general-capability suite and compared to the base model calibration. Overconfidence in out-of-distribution queries has been characterized.
9. The run manifest is stored as an immutable artifact alongside the model checkpoint and includes dataset version, base model version, training configuration, and random seeds.

10. Catastrophic forgetting has been explicitly checked using the general-capability suite. Mitigation measures (PEFT, learning rate reduction, rehearsal) are documented and applied.
11. Multi-adapter serving has been evaluated against the serving workload. Adapter cache hit rates are measured and meet the latency budget requirements.
12. Per-tenant adapter routing isolation has been verified: audit logging confirms that each request received the correct adapter and no cross-tenant contamination path exists.
13. A rollback plan exists with a named owner, a specified prior version to revert to, a target time-to-rollback, and a documented procedure that has been rehearsed at least once.
14. A re-evaluation schedule is recorded at ship time, with a next review date, the metrics that will trigger deprecation, and the deprecation notification procedure for consuming teams.
15. Adapter artifacts, run manifests, and evaluation results are archived under a retention policy that satisfies the compliance obligations for the data used in training.

6.15 Team Conversation

Use these questions in team meetings or design reviews to pressure-test alignment on fine-tuning and adaptation discipline. They are designed to surface assumptions and gaps before they manifest as production incidents.

1. When did we last decide to fine-tune, and if we repeated that evaluation today using the current base model's capability baseline, would we make the same call? What has changed in the model landscape since that decision?
2. Could we reproduce our most recent successful adapter from scratch, using only the artifacts stored in our experiment

tracking system and dataset registry? If not, what is missing, and what class of incident would that gap cause?

3. What does our general-capability held-out suite look like, and when did it last catch a regression that was not visible in the in-distribution test set? If it has never caught a regression, is the suite actually testing general capability, or is it implicitly correlated with the training distribution?
4. Where is catastrophic forgetting most likely hiding in our currently deployed adapters? Which adapters have not been evaluated on a general-capability suite since they shipped?
5. Is our adapter strategy one model per tenant or one model with many adapters, and is that strategy the right match for our current workload size and growth trajectory?
6. Have we ever executed an adapter rollback cleanly under production pressure? If yes, what made it work? If not, has the rollback plan been rehearsed in a staging environment?
7. Which of our current fine-tuning use cases would we now solve with retrieval-augmented generation or improved prompt engineering, given today's model capabilities? How often do we revisit that question?
8. How do we handle adapter retirement today? Is there a registry that tracks which consuming systems depend on each adapter version, and does it drive automated deprecation notices?
9. What is our inter-annotator agreement on the labeling guidelines for our most recent instruction tuning dataset, and how does that agreement level correlate with the model's behavioral consistency on edge cases?
10. If the base model version our adapters depend on becomes unsupported in six months, what is the upgrade path, and who owns it?

6.16 Key Takeaway

Adaptation is a discipline, not a one-shot action. The teams that produce reliable adapted models in production do not succeed because they have a better training run; they succeed because they have built the scaffolding around the training run that makes it interpretable, reproducible, and safe to operate over time. That scaffolding has five load-bearing elements: an explicit decision process that evaluates fine-tuning against prompting and retrieval before committing to a training run; a data curation pipeline that treats the dataset as a versioned, auditable artifact; a dual evaluation framework that separates in-distribution success from general-capability regression; a reproducible pipeline that captures every input needed to replay the run; and a deprecation plan written at ship time that ensures the adapter is retired as deliberately as it was launched.

Parameter-efficient fine-tuning is the right default for most teams operating at production scale. It reduces compute cost, preserves the base model's general capabilities by construction, and enables multi-adapter serving architectures that are impractical with full fine-tuning. Full fine-tuning remains the right choice for a narrow set of scenarios where the required capability shift genuinely exceeds what PEFT can deliver. Still, those scenarios are rarer than most teams assume. The first question before any adaptation run should be whether a higher-quality prompt can solve the task, and the second should be whether the gap is a knowledge gap better served by retrieval. Fine-tuning earns its complexity only when both prior options have been explicitly evaluated and found insufficient.

Data quality is the dominant lever in adaptation outcomes, and that lever is fully under the team's control. Compute budget, model architecture, and hyperparameter choices all matter at the margin, but the quality and coverage of the training data almost always set the ceiling on adaptation quality. Teams that invest in deduplication, decontamination, balanced sampling, and consistent labeling guidelines before their first training run consistently outperform teams that iterate on training configuration without addressing data quality. Adapters need a deprecation plan from the day they ship; the operational clarity that comes from knowing which adapters are active, which are scheduled for review, and which are retired is the pilot-to-production pathway that separates responsible-by-design adaptation from the accumulation of technical debt that eventually requires an emergency cleanup.

7 Prompt Engineering and Prompt Ops

7.1 Opening Scenario

The page arrived at 8:47 on a Monday morning. A senior engineer at a mid-size financial services firm had pushed a prompt change the previous Friday afternoon, just before logging off for the weekend. The change looked harmless in the diff: a single sentence in the system prompt had been reworded to make the assistant sound more helpful. No reviewer flagged it. There was no automated evaluation gate. The change went live within minutes of the merge.

By Monday, the customer support team had escalated two hundred tickets. The assistant was recommending account actions that fell outside its approved scope, including transfers and dispute filings that required human authorization. More quietly, the data team had caught a billing spike: average tokens per response had doubled overnight, doubling the day's inference cost with a single sentence change. Neither failure would have been obvious from the diff. Both were obvious in production.

The post-incident review surfaced a pattern familiar to any team that has grown an LLM system without operational discipline. Prompts were scattered across source files, configuration databases, a vendor console, and an engineer's personal notes. Nobody could answer a basic audit question: which prompt version was running at 3:00 Friday afternoon? The fix required pulling logs from three systems and piecing together a timeline manually.

The root cause was not a careless engineer. It was an operating model that treated prompts as configuration text rather than as production artifacts. The team used version control for their application code, a CI pipeline with automated tests, and a change-approval process for schema migrations—none of those disciplines applied to the prompts. The Friday afternoon change was not a mistake; it was the expected outcome of a system with no evaluation gate, no rollback capability, and no governance model for the highest-edit surface in the LLM stack.

This chapter is the playbook that closes the gap. It covers prompt design as an engineering discipline, composition patterns that scale, structured-output contracts that keep LLMs cleanly integrated with the rest of the stack, testing and versioning strategies that make changes auditable, and governance and cost patterns that turn prompt ops into a peer of DevOps and MLOps rather than an afterthought.

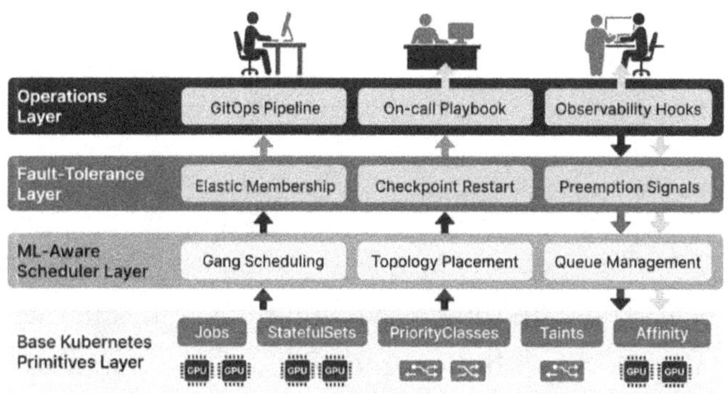

7.2 Why It Matters

Prompts are the most-edited surface in a production LLM system and, for most teams, the least-managed one. A model weight change requires a retraining run, a review cycle, and a deployment gate. A prompt change can be live in seconds, affecting every user the system touches. The asymmetry is not hypothetical. Teams that treat prompts as configuration text discover the gap the hard way: a changed sentence doubles inference cost, shifts the assistant's behavior outside its approved scope, or introduces a regression that takes days to isolate because there is no version history to diff.

The workflow-level impact of unmanaged prompts extends beyond individual incidents. When prompts are not versioned, debugging production behavior requires archaeology rather than inspection. When prompts are not tested, every change is a gamble rather than a controlled experiment. When prompts are not governed, the boundary between what the assistant is authorized to do and what it will do under novel phrasing is determined by whoever last edited the prompt. In a regulated industry, that gap is not an engineering oversight; it is a compliance risk that regulators increasingly know to look for.

Building prompt ops as a first-class engineering discipline is a mission-aligned investment. The teams that do it create a pilot-to-production pathway where every prompt change is auditable, every behavioral regression is caught before it reaches users, and the cost impact of any single edit is bounded by policy rather than discovered in the billing dashboard. The patterns in this chapter are not theoretical. They are the operational floor that separates a team that operates LLMs with the same confidence it operates the rest of its stack from a team that hopes nothing goes wrong over the weekend.

> *This chapter and Chapter 9 divide the work of prompt quality between them, and a reader looking for the wrong half here will not find it. Chapter 4 owns prompt design, the prompt lifecycle, versioning policy, prompt governance and review workflow, and the prompt*

operations maturity model: the question of how a fleet of prompts is designed, changed, approved, and managed over time. Chapter 9 owns the concrete machinery of verifying that a prompt is correct: the unit tests, the eval suites with golden sets, the LLM-as-judge patterns, and the continuous-integration gates that block a regression before it ships. A reader looking for how to write a prompt test belongs in Chapter 9. A reader looking for how to govern prompt changes across a fleet stays in Chapter 4.

7.3 Prompt Design Patterns and Templates

A prompt is not free text handed to a model and hoped to produce useful output. In production, a prompt is a structured artifact with named parts, each carrying a different kind of authority and a different failure mode. The three-part structure that has stabilized across most production deployments separates the system instruction, the conversation context, and the user input into distinct zones. The system instruction establishes the model's role, scope, tone, and hard constraints. The conversation context carries retrieved documents, tool outputs, or prior turns. The user input is the live request that the system cannot fully control. Keeping these zones explicit is the first discipline of production prompt design.

Role prompting is the most widely used design pattern and the most often misunderstood. Declaring a role for the model, such as a financial advisor, a code reviewer, or a customer support agent, does not restrict the model to that role unless the system instruction also defines what the role excludes. A system prompt that says 'You are a helpful financial advisor' without scope constraints will answer questions about tax law, estate planning, and investment strategy with equal confidence, regardless of whether those topics are within the intended deployment scope. Production role prompts include both a role declaration and an explicit scope boundary: what the assistant will do and what it will decline to do.

Structured task prompts are the next step above role prompts. They encode not just the model's identity but the expected format of its response, the steps it should take before answering, and the conditions under which it should decline rather than attempt an answer. A structured task prompt for a document summarization assistant might specify the output length, require the model to identify the document type before summarizing, and instruct it to return a specific refusal phrase when the document does not meet quality thresholds. Each of those specifications reduces the variance of the model's output and makes the surrounding application code more predictable.

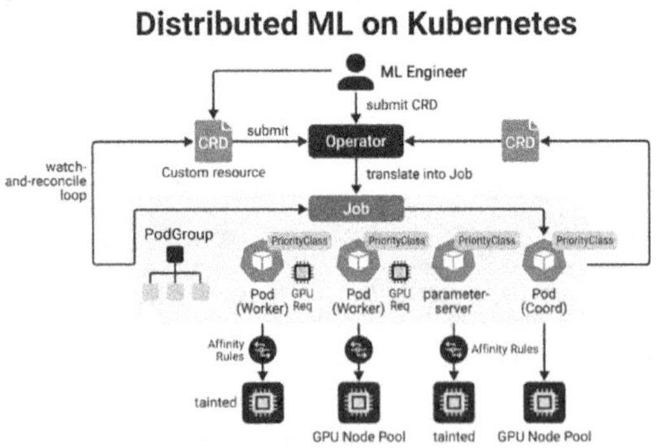

Template patterns make prompts composable and reduce the number of distinct prompt strings a team must maintain. A template separates the fixed instruction structure from the variable slots that change per request. A code review template might address the review criteria, output format, and escalation language while leaving slots for the language, the code block, and the reviewer's focus area. Templates are version-controlled as first-class artifacts: the template version is part of the prompt's identity, and a change to the template triggers the same evaluation and review process as a change to a specific prompt.

Few-shot examples are a template element that deserves its own operational discipline. Adding examples to a prompt improves model performance on structured tasks. Still, examples also increase token consumption, introduce latent bias if the examples are not representative, and can become stale as the target task evolves. Production few-shot libraries are maintained as separate artifacts from the prompts that reference them, tagged with the task type and the model version they were calibrated for, and reviewed when either the task requirements or the underlying model changes.

7.4 Prompt Chaining and Composition

Single-turn prompts handle a narrow class of production tasks. Most real workflows involve multiple steps: a first call to extract structured data from unstructured input, a second call to validate the extraction, and a third call to generate a response using the validated data. Prompt chaining is the pattern that connects these steps, and it introduces failure modes that do not exist in single-turn systems. Understanding those failure modes is a prerequisite for building chains that behave predictably under production load.

The most common chain failure is error amplification. If the first step in a chain produces a subtly wrong output, every downstream step operates on that flawed intermediate. In a single-turn system, a bad output is visible immediately. In a chain, a bad output from step one may pass through two or three downstream steps before the final output reveals something is wrong, at which point the causal trail is difficult to reconstruct without logging the intermediate outputs of every step. Production chains always log intermediate outputs as named artifacts, not just the final result.

Router patterns are a chain variant that sends requests to different sub-chains based on a classification step. A customer support system might classify incoming requests as billing questions, technical issues, or account changes, and route each class to a specialized prompt chain tuned for that domain. Routers reduce the risk of a single general-purpose prompt drifting in response to adversarial or edge-case inputs, because each sub-chain handles a narrower scope. The router itself is a prompt that must be tested and versioned with the same rigor as any other production artifact.

MPI Orchestration Pattern on Kubernetes

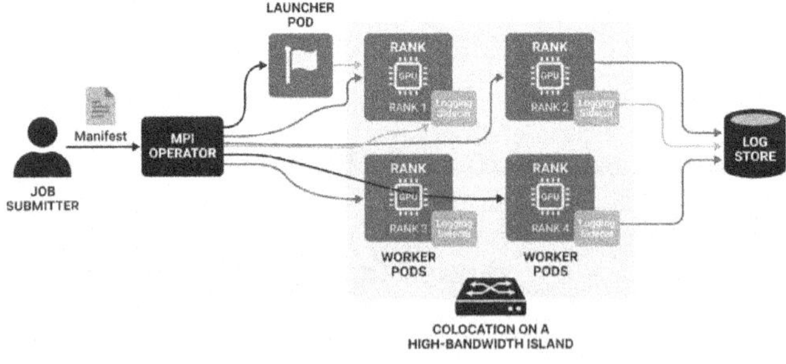

Orchestration patterns govern how chains handle partial failures, retries, and timeouts. A chain without an error-handling policy will either surface internal model errors directly to users or silently return incomplete results. The discipline is to define, for each step in every chain, the expected output shape, retry policy, timeout budget, and fallback behavior. Fallback behavior for a content generation step might be to return a cached prior response. Fallback behavior for an extraction step might be to return a structured null that downstream steps know how to handle. These policies are encoded in the orchestration layer, not in the prompts themselves, so they can be changed without touching prompt content.

Chain composition creates a surface area for prompt injection that does not exist in single-turn systems. When the output of one prompt becomes part of the input to another, an adversarially crafted output from step one can attempt to redirect step two. A document extraction step that returns text from an untrusted source, for example, can embed instructions in that text that the next step's prompt treats as directives. Defending against this requires treating every external input as untrusted, passing structured data between chain steps rather than raw text where possible, and validating intermediate outputs against expected schemas before passing them forward.

7.5 Structured Outputs, JSON Mode, and Function Calling

Structured outputs are the primary integration interface between LLMs and the rest of the production stack. When a model returns a response in a known schema, the application code consuming that response can be written against the schema rather than against prose. Schema-driven integration is more reliable, more testable, and easier to monitor than parsing unstructured text. The discipline of requiring structured outputs wherever the downstream system needs to act on the model's response is one of the highest-leverage practices in production prompt engineering.

JSON mode and constrained decoding are the two main mechanisms for enforcing structured outputs at the model level. JSON mode instructs the model to produce valid JSON as its entire output. Constrained decoding goes further: it uses the model's token generation process to guarantee that the output conforms to a specified schema, rejecting tokens that would violate it before they are emitted. Constrained decoding produces harder guarantees than JSON mode but requires model support and adds latency. JSON mode is widely supported and appropriate for most production use cases when combined with schema validation on the receiving side.

Function calling is a structured-output pattern that models a response as a call to a named function with typed arguments. The model does not execute the function; it returns a structured specification of which function to call and with what arguments. The application layer validates the specification against the function's signature and executes the call. Function calling is the standard pattern for tool-using LLM systems because it separates the model's intent, expressed as a structured specification, from its execution, which occurs in verified application code with normal error handling and access control.

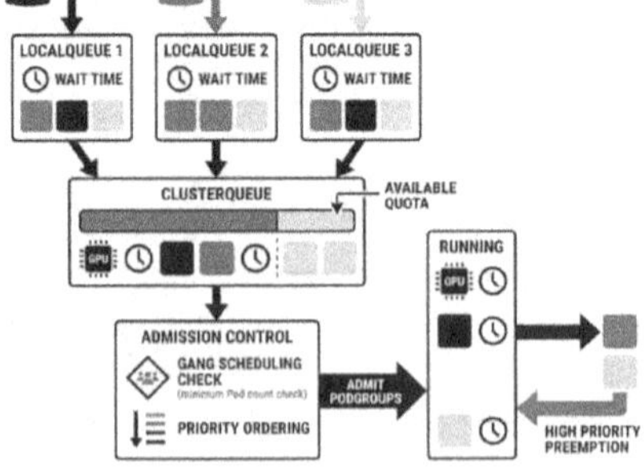

Schema design for structured outputs is an engineering task that deserves the same care as API design. A schema that is too permissive (accepting any string for a field that should be an enumeration) provides little safety and complicates downstream validation. A schema that is too strict (requiring exact string matches for fields that the model naturally expresses with slight wording variations) leads to high failure rates and forces application code to work around the validation. Production schemas are designed iteratively: start with the minimum structure the application requires, run evaluation against a representative input set, and tighten the schema based on observed failure modes.

Validation is not optional for structured-output integrations. A model configured for JSON mode can still produce invalid JSON under novel inputs. A model configured for function calling can produce a valid JSON structure that references a function name that does not exist or passes arguments outside the defined type constraints. Every structured output that crosses a system boundary must be validated against its schema before the application acts on it. Validation failures are logged with the full prompt context and reported as a distinct failure class in the system's observability layer, separate from model errors and application errors.

7.6 Prompt Testing and Versioning Strategies

Testing prompts requires a different mental model than testing deterministic code. A unit test for a function asserts that a specific input produces an exact output. A unit test for a prompt asserts that a specific input produces an output that satisfies a set of quality criteria, because the model's output varies across runs. The testing discipline for prompts builds around golden-input suites, rubric-based evaluation, and regression detection rather than exact-match assertions.

Golden-input suites are a collection of representative inputs paired with expected output properties. A golden input for a summarization prompt might specify that the output should be between 100 and 200 words, should mention the three key entities from the source document, and should not introduce information not present in the source. Each of those criteria is a test assertion, evaluated by a checker that may be a rule-based function, a regex, or a secondary LLM call. The team curates the suite to cover the happy path, known edge cases, and the specific failure modes the prompt has exhibited in production.

Regression suites run on every prompt change before merge. The regression test compares the current prompt's output on the golden suite against the baseline established by the previous prompt version. A regression is detected when the pass rate on any criterion falls below a configured threshold. The threshold is set by the team based on the task's sensitivity: a customer-facing response-generation prompt might require that no regression exceed two percent on any criterion. In comparison, an internal classification prompt might accept five percent regression on secondary criteria if primary accuracy improves.

Topology-Aware Placement: Rack, Switch, and NUMA Hierarchy

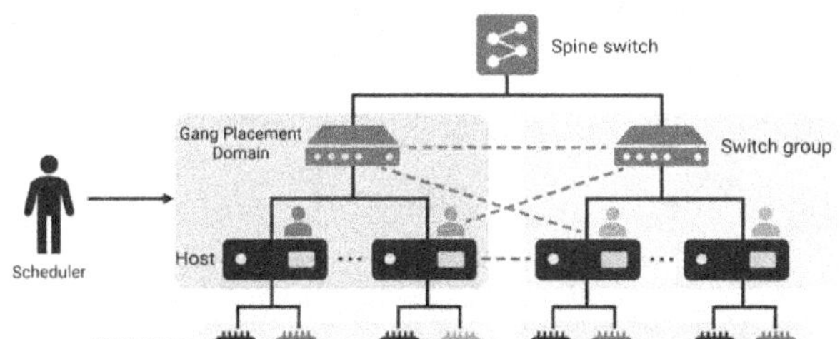

Versioning prompts require treating them as software artifacts rather than configuration values. Each prompt version has a unique identifier, a change log entry explaining what changed and why, and a link to the evaluation results that justify the change. The versioning scheme must support rollback: if a newly deployed prompt version causes an unexpected behavioral shift in production, the team should be able to identify the prior version, restore it in a single operation, and be confident that the rollback returns behavior to a known state.

Prompt version identifiers should be semantic rather than sequential. A convention that encodes the major behavioral intent, the iteration within that intent, and the date of the change is more informative than an auto-incremented integer. When a production incident requires determining which prompt version was running at a specific timestamp, a semantic version identifier provides enough context to be useful in the incident timeline without requiring a lookup table.

Deployment gates for prompt changes mirror the gates used for application code. A pre-merge gate runs the regression suite and fails the change if any criterion regresses beyond the threshold. A post-deploy gate runs a subset of the evaluation suite against the live system on a canary population. It triggers an automatic rollback if the live pass rate deviates from the pre-deploy baseline. The canary gate is the last line of defense before a prompt change reaches full production traffic, and it is the gate that the Friday afternoon scenario lacked.

7.7 Eval-Driven Development for Prompt Changes

Eval-driven development is the prompt engineering analog of test-driven development. The discipline is to write the evaluation criteria before writing the prompt change, then use the evaluator's judgment to guide the prompt revision, and ship only when the evaluation score improves relative to the baseline. The practical effect is that every prompt change has a defined success criterion before any prompt text is written, and the success criterion is measured objectively rather than subjectively.

The workflow starts with the team agreeing on what the prompt change is trying to improve. If the goal is to reduce the rate at which the assistant recommends out-of-scope actions, the team writes a test set of inputs that should trigger refusals under the new prompt and a test set that should be handled normally. They define the evaluation rubric: the proportion of refusal inputs correctly declined, the proportion of normal inputs answered correctly, and the threshold improvement needed to justify the change. Only then does anyone write prompt text.

LLM-as-judge evaluation is an important tool in the eval-driven workflow, but it requires controls in place. Using a secondary model to evaluate the primary model's output introduces its own error rate and biases. The secondary model may have preferences that do not align with the task's actual quality criteria. The discipline is to calibrate the judge: run the judge on a human-labeled set of outputs and measure its agreement rate with human judgment. A judge who agrees with human raters on a rubric criterion at 85 percent is a credible evaluator for that criterion. A judge with 60 percent agreement is providing noise rather than a signal.

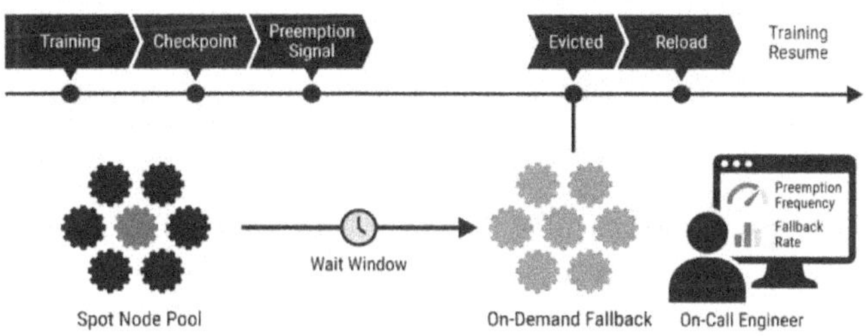

Eval coverage as a release-gate metric treats the breadth and depth of the evaluation suite as a first-class health signal, not just the pass rate on existing tests. A prompt that passes all twenty tests in a sparse evaluation suite may still exhibit unexpected behavior on the long tail of production inputs the suite does not cover. Teams that invest in expanding their eval suites over time, adding new test cases whenever a production incident reveals a gap, build a progressively stronger signal about their prompts' behavioral boundaries. Eval coverage is tracked in the same dashboard as pass rate and regression rate.

The distinction between pre-merge evaluation and post-deploy evaluation is operationally important. Pre-merge evaluation runs on a static test set and validates that the change improves known inputs. Post-deploy evaluation runs on live traffic samples and validates that the improvement holds on the full distribution of real requests, including edge cases the test set did not anticipate. A complete eval-driven workflow requires both gates, with clearly defined thresholds for each and an automatic rollback trigger on the post-deploy gate.

7.8 Prompt Governance and Review Workflows

Governance distinguishes between prompts that require the same review process as application code and prompts that require a higher-stakes approval workflow. Not all prompts carry the same risk profile. A prompt that controls the formatting of an internal report summary has a low blast radius if it behaves unexpectedly. A prompt that controls what actions a customer-facing assistant can recommend has a high blast radius, and its change process should reflect that. Governance frameworks for prompts classify them by risk tier and apply proportional review requirements. Architects own the boundary decisions: which template families exist, who can author them, where they live in the artifact registry, and how they relate to the rest of the platform's policy surface.

High-stakes prompt changes, those that affect customer-facing behavior, involve scope constraints, or carry regulatory implications, require a multi-party review before deployment. The review board for a high-stakes prompt change includes the engineering owner. This product representative can confirm the intended behavior, a compliance representative can assess regulatory implications, and, where the model's behavior involves safety-relevant domains, a safety reviewer can assess safety implications. The review is not a rubber stamp; it includes running the evaluation suite together and explicitly signing off on the pass rate before the change is promoted to production.

Owner assignment is the operational backbone of prompt governance. Every production prompt has a named owner accountable for its behavior, evaluation coverage, and lifecycle. Owner assignment prevents the common failure mode of collective ownership, which in practice means no ownership: nobody maintains the evaluation suite, nobody reviews the change log, and nobody is paged when the prompt's behavior drifts. A prompt without a named owner is a prompt waiting to cause an incident.

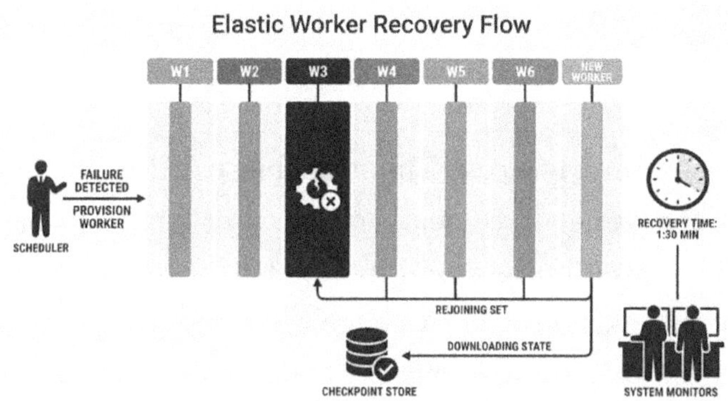

Audit trail requirements add a documentation discipline to the governance framework. An audit trail for a prompt captures, at minimum, the prompt version identifier, the deployment date, the name of the reviewer who approved it, the evaluation suite pass rates at the time of approval, and any exceptions or conditions attached to the approval. Teams in regulated industries must be able to reconstruct this trail for any prompt currently in production and for any prompt that was in production at a specified past date. Storing the audit trail in a system designed for it, rather than in commit messages and shared documents, is the difference between a governance process that survives an audit and one that collapses under scrutiny.

Approval workflows should be automated at the mechanical level and handled by humans at the judgment level. Automation handles classification: a change to a prompt tagged as Tier 1 automatically routes to the full review board workflow without requiring the engineer to remember to do so. Automation handles scheduling: the review board convenes within a defined SLA for Tier 1 changes, and the change expires and requires resubmission if the SLA passes without a decision. Human judgment handles the substance: the review board evaluates the change against its criteria and approves or rejects it based on the evaluation results, not on the diff alone.

7.9 Cost-Aware Prompting Strategies

Every token in a prompt is a cost. The system instruction, the few-shot examples, the retrieved context, and the user input all count toward the input token budget, and the model's response counts toward the output token budget. An engineer who writes a 2,000-token system prompt for a task that a 300-token prompt handles equally well is spending six times more per request without measuring the return. Cost-aware prompting starts with measuring the token cost of each prompt component and asking, for each component, whether a measurable quality improvement justifies the cost.

Instruction conciseness is the first cost lever. System prompts that repeat the same constraint in three different phrasings, include extensive background context the model does not use, or embed formatted examples when unformatted examples work equally well, are common in production systems. A conciseness audit reviews each sentence in a system prompt against the evaluation suite: Does removing this sentence reduce the pass rate? If not, the sentence does not belong in the production prompt. The audit is not a one-time exercise; it runs as part of every major prompt revision.

Model tiering by task is a structural cost strategy that pairs each task type with the lowest-capability model that meets the quality bar for that task. A production system that uses a single frontier model for every task, from high-stakes reasoning to simple extraction, is almost always overspending on the simple tasks. The pattern is to identify the quality threshold for each task class using the evaluation suite, run a set of candidate models against that threshold, and route each task class to the cheapest model that meets it. The routing logic is part of the prompt ops infrastructure, not part of any individual prompt.

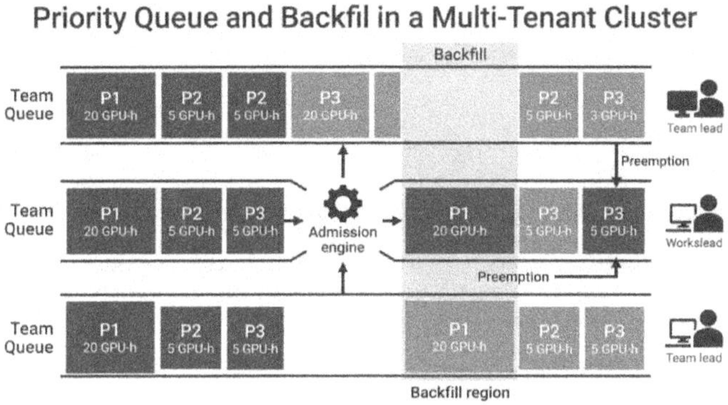

Priority Queue and Backfil in a Multi-Tenant Cluster

Retrieval-as-context budget is a pattern that replaces static context blocks in system prompts with dynamically retrieved, task-relevant context. Instead of embedding a 1,500-token product catalog excerpt in every system prompt, the prompt calls a retrieval step that returns only the three or four catalog entries most relevant to the current user request. The total context is smaller per request, the relevant information is denser, and the model's attention is focused on what matters for the current task rather than diluted across a large static block. The retrieval step adds latency, so the pattern applies where the context block is large and the per-request relevance is highly variable.

Worst-case token spend is a policy, not a discovery. Teams that bound worst-case token spend explicitly, by setting maximum input and output token limits per request class in the orchestration layer, can reason about cost exposure before a traffic spike rather than after. A prompt that can theoretically receive an unbounded user input is a cost exposure that compounds with traffic. The policy is to define the maximum input length the application will pass to the model, truncate or summarize inputs that exceed that length before passing them, and configure output token limits that match the task's actual output range rather than the model's maximum.

7.10 Prompt and Response Caching

Caching bends the cost-and-latency curve more than most teams realize. A well-positioned cache layer can eliminate most model inference calls for workloads in which a significant fraction of requests are semantically equivalent, returning precomputed responses instantly and at negligible cost. The challenge is deploying caching patterns that are safe: they do not return stale responses for inputs whose correct answer has changed, and they do not collapse distinct inputs into the same cached response when their differences are semantically meaningful.

Exact-match caching stores the model's response to a specific prompt string and returns it for any subsequent request with the identical string. It is the simplest caching pattern and has the strongest correctness guarantee: two requests that are character-for-character identical will always have the same correct response if the underlying information has not changed. Exact-match caching is highly effective for fixed system prompts, for FAQ-style applications where the same questions recur frequently, and for batch processing jobs where the same prompt is submitted many times with the same content.

Semantic caching extends caching to paraphrased prompts rather than exact matches—the cache stores an embedding of each prompt, along with its response. For a new request, the cache computes the embedding of the incoming prompt and retrieves any cached entry within a configured cosine similarity threshold. If a sufficiently similar entry exists, the cached response is returned without a model call. Semantic caching significantly increases cache hit rates for applications in which users phrase the same underlying question in different ways. Still, it introduces a correctness risk: two semantically similar prompts may not have identical correct answers. The similarity threshold must be calibrated conservatively, and semantic cache entries must carry metadata that allows the system to invalidate them when the information they encode changes.

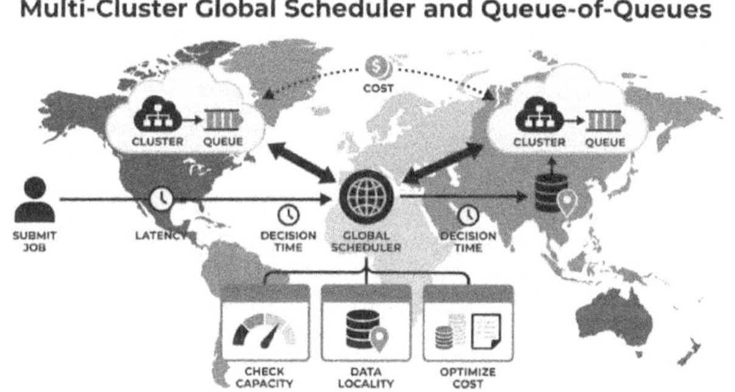

Prefix caching is a model-level optimization that most inference providers now support. When multiple requests share a long common prefix, such as a fixed system prompt followed by different user inputs, the provider can cache the KV (key-value) state computed for the shared prefix and reuse it across requests rather than recomputing it from scratch. The practical effect is a reduction in both latency and cost for requests with long system prompts, because the computational cost of processing the fixed prefix is paid once and amortized across all requests that share it. Prompt design should account for prefix caching: placing stable content at the beginning of the prompt and variable content at the end maximizes the fraction of the prompt that qualifies for prefix caching.

Cache consistency and eviction policies determine whether a cached response is still correct when it is returned. Responses that encode time-sensitive information, such as pricing, availability, or regulatory guidance, have a natural staleness horizon beyond which they should not be served from cache. Eviction policies for these response classes set an explicit TTL (time-to-live) that matches the information's natural update cadence. Responses that encode structural knowledge with low update frequency can carry longer TTLs. The team should be able to answer the question of what the worst-case staleness of any cached response class is, and that answer should be driven by business requirements rather than by the caching framework's default behavior.

7.11 Multi-Modal and Multi-Lingual Prompt Patterns

Multi-modal prompts introduce inputs beyond text: images, audio, structured documents, and video frames are all inputs that current models can reason over alongside text instructions. The prompt engineering discipline for multi-modal inputs must account for the fact that the model's behavior on combined text-and-image inputs is not simply the sum of its text and image behaviors. Models exhibit different capability profiles on multi-modal inputs, and those profiles must be evaluated independently using multi-modal test sets rather than inferred from text-only evaluations.

Image inputs in production prompts require explicit handling of image quality, resolution, and content constraints. A prompt that works correctly on high-resolution, well-framed photographs may fail on the compressed, cropped, or low-light images that real users submit. The evaluation suite for an image-input prompt must include samples that reflect the actual distribution of image quality in production, not just the curated examples used during development. Input validation for image prompts should reject images below a configurable quality threshold before the prompt is submitted to the model, with an explicit error message rather than a silently degraded response.

Cross-language prompting is a distinct challenge from multi-modal prompting. A system prompt written in English may produce correct behavior for English user inputs and subtly different behavior for inputs in other languages, because the model's instruction-following behavior and its safety guardrails have been evaluated more thoroughly in English than in other languages. Teams that deploy to multi-lingual user populations must evaluate their prompts separately for each supported language, using native speakers to curate the evaluation set rather than machine-translated versions of the English test cases.

GitOps Pipeline for Machine Learning Job Definitions

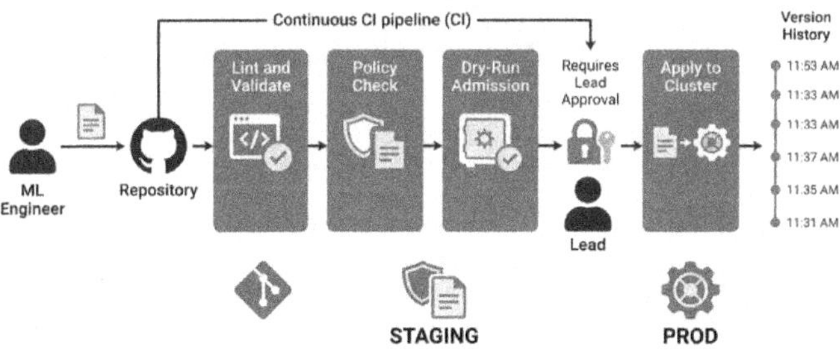

Localization patterns prevent a prompt that performs well in one language from silently regressing in another. The pattern is to maintain a localization baseline for each supported language: a version of the evaluation suite translated and verified by a native speaker, run against the current prompt version, with a documented pass rate. When a prompt change is proposed, the localization suite runs for all supported languages, not just the development language. A change that improves the English pass rate at the cost of the French or Spanish pass rate is not an improvement; it is a regression that affected users will discover before the team does.

Audio inputs carry additional operational considerations beyond transcription accuracy. A voice-input prompt that receives transcription artifacts, background noise transcribed as content, disfluencies, or speaker overlap must handle those inputs gracefully rather than producing responses that are correct for a perfect transcript but incorrect for the actual input. The evaluation suite for audio-input prompts includes synthetic corruption of the transcription to test robustness to common transcription errors.

7.12 Prompt Lifecycle Management

Prompts are born in draft, promoted to staging, released to production, and eventually retired. Each stage of that lifecycle has its own requirements. Teams that skip lifecycle discipline discover the consequences at production scale: prompts that were deployed for a specific use case continue to run after the use case has changed, prompts that were effective for an earlier model version degrade silently after a model upgrade, and prompts written for a single user type accumulate behavior that serves no current user well.

Draft prompts live in a development environment where evaluation runs, but no real user traffic is served. The draft phase is where the prompt author iterates against the evaluation suite, using the eval-driven workflow to establish a baseline that justifies promotion. A draft prompt that does not meet the evaluation suite's minimum threshold is not promoted to staging, regardless of how promising it appears on manual inspection. The evaluation suite is the objective arbiter.

Staging prompts run against a subset of production-like traffic, either a canary population or a synthetic replay of recent production requests, with full observability. The staging phase validates that the prompt's evaluation suite performs as expected under the real distribution of inputs, including edge cases and adversarial patterns that the development evaluation set may not have covered. The staging gate also validates operational concerns, including latency at the 95th and 99th percentiles, token consumption distribution, cache hit rate, and cost per request.

GPU and NUMA Topology-Aware Resource Scheduling

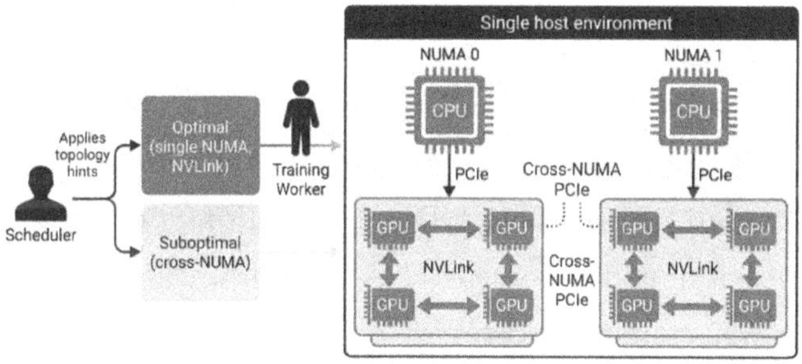

The deprecation policy defines how a prompt moves from active production use to retirement. A prompt enters deprecation when a new prompt version replaces its use case, when the model it was calibrated for is retired, or when the team has determined that the prompt no longer serves its intended purpose. Deprecated prompts are marked read-only in the prompt registry: no further changes are made to them, and the documentation is updated to reference the replacement. Downstream systems that reference the deprecated prompt receive a migration notice with a timeline.

Retirement archives the prompt and its full history: every version, every evaluation result, every change log entry, and the audit record of its production deployments. The archive is not deleted; it is stored in a read-only repository that supports lookup by prompt identifier and by deployment date. Regulatory requirements in some industries specify retention periods for model artifacts; the prompt archive must satisfy those requirements. A team that can say that it can produce the exact prompt text, evaluation results, and deployment history for any prompt that was in production at any date in the past two years has achieved the operational clarity that prompt lifecycle management is designed to enable.

Documentation standards for production prompts turn tribal knowledge into auditable assets. A well-documented prompt includes the use case it serves, the model and version it was calibrated for, the evaluation suite it passes, the owner responsible for maintaining it, the date it was deployed, and the conditions under which it should be reviewed. Documentation is maintained in the same version control system as the prompt text itself, updated with every change, and reviewed as part of every change approval. A prompt without documentation is a liability in production because it cannot be safely modified, retired, or audited.

7.13 Prompt Ops Maturity Model

The prompt ops maturity model gives teams an honest framework for assessing their capabilities and identifying the next stage to invest in. Like software engineering maturity models, it describes five stages: ad hoc, repeatable, defined, managed, and optimizing.

Teams rarely advance uniformly across all dimensions; a team might be at the defined stage for versioning and the ad hoc stage for cost management. The model's value is as a diagnostic tool, not a progress report.

At the ad hoc stage, prompts have no formal versioning, an evaluation suite, or a governance process. Changes are made directly in production or in a configuration database with no change log. The team knows approximately which prompts exist, but cannot reconstruct their history. Cost per request is not tracked at the prompt level. This stage is where most teams start and where many teams remain longer than they should.

At the repeatable stage, prompts are in version control, each has a named owner, and a basic evaluation suite exists for the highest-stakes prompts. Changes to high-stakes prompts require a second reviewer. Cost per request is tracked at the system level. Rollback is possible but requires manual effort. At the defined stage, every production prompt has a complete evaluation suite; changes require passing the evaluation suite as a pre-merge gate; governance tiers are formalized; and cost is tracked at the prompt level. At the managed stage, the evaluation suite is comprehensive, the post-deploy canary gate is automated, prompt performance metrics feed into the team's standard operational dashboards, and cost bounds are enforced by policy in the orchestration layer. At the optimization stage, the team continuously improves its evaluation suite based on production signals, uses eval-driven development as the default workflow for all prompt changes, and employs a caching and model-tiering strategy that actively minimizes cost without degrading quality.

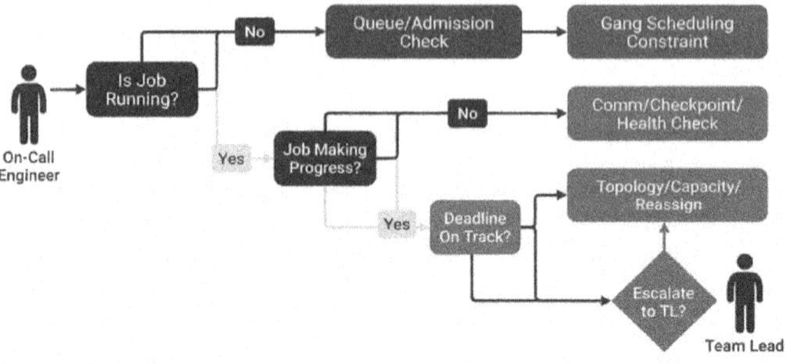

Advancing maturity stages requires investment at each transition. Moving from ad hoc to repeatable requires investment in tooling for version control and evaluation infrastructure. Moving from repeatable to defined requires the team to build evaluation suites for all production prompts, which is labor-intensive when done retroactively. Teams that build the evaluation suite alongside the prompt, as the eval-driven workflow prescribes, arrive at the defined stage without the retroactive debt. Moving from defined to managed requires investment in automation for the CI pipeline and the post-deploy canary system. Moving from managed to optimizing requires a cultural shift: the team must internalize the eval-driven workflow as the default, not as an additional step.

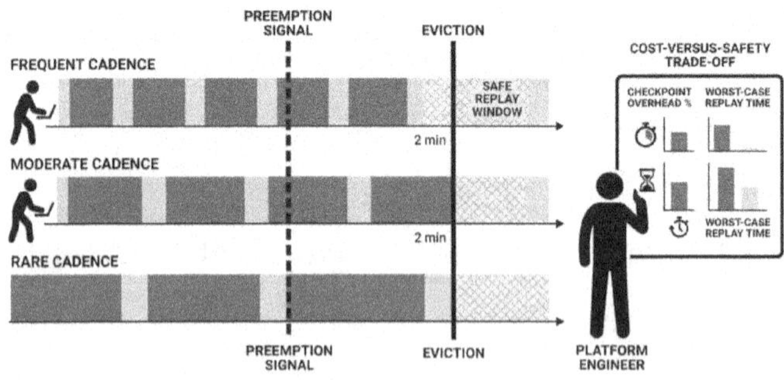

The prompt registry is the operational hub that connects all prompt ops disciplines. It stores every prompt version, its associated metadata, its evaluation results, its governance record, and its deployment history in a single queryable system. Teams that build a registry early can answer operational questions that teams without one cannot: which prompt version was running at a specific timestamp, which prompts are owned by a specific engineer who is leaving the team, which prompts have not had their evaluation suites updated in the past quarter, and which prompts are candidates for cost reduction through model tiering. The registry is the foundation that makes PromptOps scalable.

7.14 Technical Checklist

1. Every production prompt is in version control with a named owner and a change log entry for every revision.
2. Every prompt change runs a regression evaluation suite as a pre-merge gate, with documented pass-rate thresholds for each criterion.
3. Every prompt has been classified by governance tier and reviewed through the appropriate workflow for that tier within the last quarter.
4. Structured-output integrations are validated against a declared schema before the application acts on the model's response.
5. A post-deploy canary gate runs evaluation against live traffic and triggers an automatic rollback if the live pass rate falls below the pre-deploy baseline.
6. Worst-case token spend per request is bounded by an explicit policy in the orchestration layer, not discovered in the billing dashboard.
7. Prompt-and-response caching is deployed where the workload supports it, with consistency and eviction policies documented for each cache tier.

8. Model tiering routes each task class to the lowest-capability model that meets that task's quality threshold, as measured by the evaluation suite.
9. Multi-modal and multi-lingual prompts are evaluated separately with test sets appropriate to each modality and language, not assumed equivalent to the text-only English baseline.
10. Every production prompt has a deprecation and retirement record that can reconstruct its deployment history for any date in the past two years.
11. The team has placed itself on the prompt ops maturity model and identified the specific investments required to advance to the next stage.
12. Prompt documentation is maintained in version control alongside prompt text and is updated as part of every change approval.
13. An eval-driven development workflow is in place: evaluation criteria are written before prompt text, and prompts ship only when the eval score improves relative to the baseline.
14. LLM-as-judge evaluators are calibrated against human-labeled test sets, with a documented agreement rate above the team's minimum threshold.

7.15 Team Conversation

1. Where do our production prompts live today, and could we produce a complete audit of every change made to any prompt in the past ninety days?
2. Which of our last three production incidents would have been detected by an evaluation gate running on every prompt change before deployment?

3. Are we treating structured outputs as binding contracts with validated schemas, or as hints that we parse from prose and hope are correct?
4. What is our worst-case token spend per request for the five highest-traffic use cases, and would the entire team agree on those numbers?
5. Where in our current stack does caching have the highest potential impact that we have not yet captured, and what is preventing us from capturing it?
6. If a regulator asked us to explain our prompt governance process for a high-stakes customer-facing assistant, what would we say, and what would we be unable to answer?
7. Which prompts in production do we wish we could retire, and what is the actual barrier to retiring them cleanly?
8. How would we detect a prompt whose behavior has drifted over the past six months due to model updates rather than prompt changes?
9. Have we evaluated our multi-lingual prompts separately in each supported language, or are we assuming the English evaluation covers all languages?
10. Where on the prompt ops maturity model are we today, and what is the single highest-impact investment to reach the next stage in the next quarter?

7.16 Key Takeaway

Prompts are first-class production artifacts. The Friday afternoon incident that opens this chapter is not a cautionary tale about a careless engineer; it is the predictable outcome of a system that treats prompts as free text rather than as versioned, tested, governed assets. The gap between a prompt that behaves correctly and one that behaves correctly under operational discipline is the difference between a team that discovers incidents in a billing dashboard and one that catches regressions before they reach production. Every pattern in this chapter, from versioning and evaluation to governance and caching, is an engineering investment that narrows that gap.

Eval-driven development is the workflow that makes prompt ops practical rather than aspirational. When evaluation criteria are written before the prompt text, when every change ships with a measurable improvement rather than a subjective judgment, and when eval coverage is tracked as a first-class health signal, the team builds confidence in its prompts' behavioral boundaries that cannot be obtained any other way. Structured outputs are the integration discipline that gives that confidence a downstream effect: a model whose outputs are validated against a schema before any application code acts on them is a model whose errors are caught at the boundary rather than propagated through the system.

Prompt ops is a peer of DevOps and MLOps, not a subset of either. It has its own tooling requirements, its own maturity model, and its own set of failure modes that are distinct from those of application code and model training. Teams that invest in prompt ops infrastructure early, building the registry, the evaluation pipeline, the governance workflows, and the caching layers as part of the pilot-to-production pathway rather than as retroactive fixes after incidents, build systems that operate with the reliability, cost discipline, and operational clarity that production LLM workloads require. That investment is not overhead; it is the engineering floor that separates responsible deployment from a system held together by hope.

8 Serving Architectures and Latency Engineering

8.1 Opening Scenario

The week before a major shopping season, a mid-sized retail platform activated an AI shopping assistant it had been piloting for three months. Internal demos had produced sub-second first tokens. Stakeholders were impressed, budgets were approved, and the team declared the system production-ready. On launch day, real traffic arrived in waves, and the experience that had felt instant in the demo room became agonizing on the live site. Shoppers saw a spinner for eight to fifteen seconds before the first word appeared. The cluster scaled out frantically, spinning up new GPU nodes as queue depth climbed, then scaled back in midway through active user sessions because the autoscaler read average CPU utilization rather than memory pressure on the accelerators. A single model reload during scale-in dropped hundreds of in-flight requests.

The post-mortem surfaced three compounding failures. First, the team had measured the average end-to-end response time during the pilot, which looked acceptable, but had never tracked time-to-first-token (TTFT) at the 90th percentile. For a streaming assistant, TTFT is the only latency number a user consciously experiences; everything after the first token feels like natural reading pace. Second, the autoscaler was a generic HTTP-request-rate scaler that had no visibility into accelerator memory saturation. When a new user session arrived, and the KV cache for that session had to be allocated, the accelerator stalled, but the scaler saw no signal corresponding to that stall. Third, the architecture had no caching layer. Every request, including those asking the same question for the fifth time that morning, hit the model at full cost and full latency.

The fix was not a larger cluster. Adding nodes would have reduced the symptoms while preserving every root cause. The actual remedy was a serving architecture rebuilt around the right primitives: streaming token delivery with explicit TTFT budgets contractually enforced in the gateway, an accelerator-aware autoscaler that scaled on KV cache utilization and request queue depth rather than CPU, a prefix cache that eliminated redundant prefill for the system prompt shared by every session, and a semantic result cache that absorbed the long tail of near-duplicate product queries. A canary deployment process, which the team had skipped in the rush to launch, was made a structural requirement for every subsequent model change so that a flawed deployment could be caught and rolled back before it affected more than 5% of traffic. The performance profile that users see today bears no resemblance to launch day, and the compute bill per session is a third of what it was. This chapter is the catalog of the patterns that produced that outcome.

Anatomy of a Large Training Run

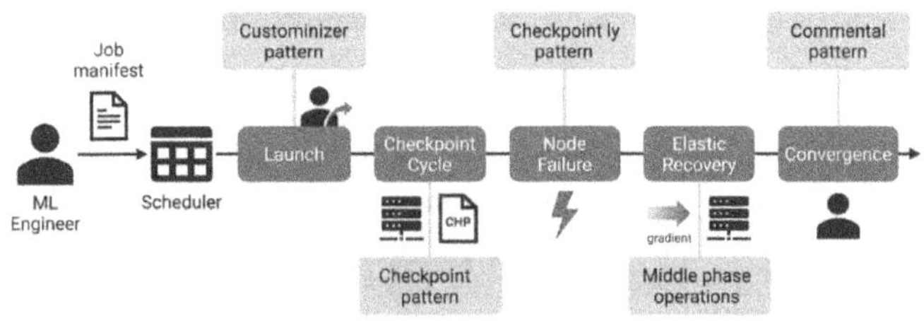

8.2 Why It Matters

Serving a large language model in production is a distinct engineering discipline. The primitives are different from those that govern classical model serving: the unit of work is an autoregressive token sequence rather than a fixed-size prediction, memory is dominated by a per-session key-value cache that grows with context length, and latency is best described not as a single scalar but as a two-part budget split between time-to-first-token and the sustained throughput that follows. Teams that import classical serving patterns without adjustment routinely encounter the same cluster of failures: tail latencies that are multiples of average latency, autoscalers that thrash during traffic spikes, and cost curves that surprise finance because no one planned for the memory-intensive prefill phase.

The workflow-level impact is greater than that of a performance tuning exercise. Latency shapes product design. A feature built around a two-second TTFT budget enables a fundamentally different user experience than one built around a twelve-second budget. Serving architecture determines which product shapes are even possible, not merely which are efficient. Cost shapes what ships. A team that has not applied compression, caching, or batching to its serving stack will find that the per-request economics make certain features commercially non-viable before they reach meaningful scale. Safe-deployment patterns determine how much risk the team absorbs with every model change; a team that cannot canary or roll back a model update is one bad deployment away from a production incident that takes down the feature entirely.

The patterns in this chapter have been extracted from the architectural decisions that separate teams operating LLMs reliably at scale from those still fighting the same fires each week. They are presented as a decision surface rather than a prescription. Each pattern names the workload properties that make it appropriate, the trade-offs it introduces, and the signals that indicate it is working. By the end of the chapter, an architect should be able to map any LLM workload onto a serving topology and justify that choice against the constraints of their organization's latency budget, cost envelope, safety posture, and deployment cadence.

8.3 Real-Time, Batch, and Streaming Serving Patterns

Every LLM workload sits on a spectrum defined by two axes: how quickly the caller needs a response, and whether the caller needs tokens as they are generated or only after generation is complete. The intersection of those axes determines the appropriate serving mode, and choosing the wrong mode often explains why a system that performs acceptably on a benchmark fails visibly in production.

Real-time synchronous serving is the pattern most teams reach for first. A caller submits a request and blocks until the full response is returned. For short outputs (a classification label, a one-sentence summary, a boolean decision), this mode is appropriate; the generation latency is low enough that the blocking wait does not degrade the user experience. For longer outputs, synchronous serving creates a problem: the caller holds an HTTP connection open for the entire generation duration, load balancers apply their own timeout policies, and users see nothing until the last token is written. The technical symptom is a sharp drop in user engagement at the point where response latency exceeds a threshold, typically around 3 seconds for consumer-facing features.

Streaming token delivery solves the user-experience problem for longer outputs. The server sends each token over a persistent connection as it is generated, typically via server-sent events or a WebSocket, and the client renders tokens as they arrive. From the user's perspective, the system starts responding immediately, and the reading pace matches the generation pace. From the engineering perspective, streaming shifts the critical metric from total response time to time-to-first-token: the latency of the prefill phase that produces the initial token becomes the primary user-facing cost, and the per-token generation rate thereafter becomes a throughput constraint rather than a latency one. Teams that have not instrumented TTFT separately from total response time often miss this shift entirely.

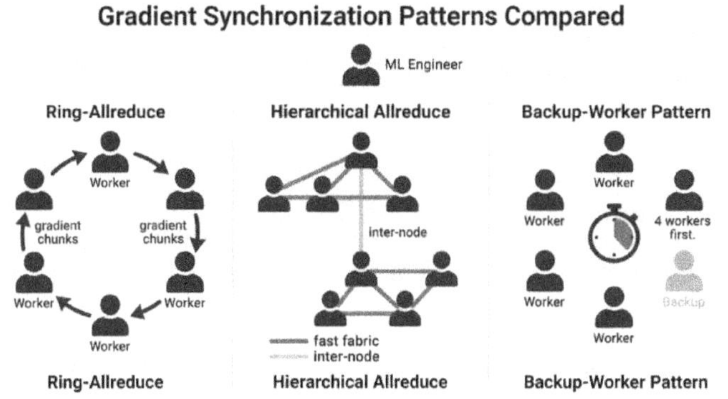

Batch and asynchronous serving patterns apply to workloads where the caller does not need an immediate response. Document summarization pipelines, overnight report generation, bulk annotation, and synthetic data production are canonical examples. In batch mode, requests are queued and processed at high throughput during off-peak hours or when dedicated capacity is available. The economic incentive is substantial: batch workloads can be co-scheduled with other jobs on shared accelerator capacity, they can use spot or preemptible instances that are unavailable for latency-sensitive traffic, and they benefit maximally from continuous batching because the system controls the scheduling of each request rather than accepting them at unpredictable arrival rates.

Asynchronous serving is the bridge pattern for workloads that require near-real-time delivery but where the caller can tolerate a callback or polling rather than a synchronous response. A user submits a long-form generation job, receives a job identifier immediately, and polls for results or receives a webhook callback when generation completes. This pattern decouples the caller's experience from generation latency and removes the need for long-lived HTTP connections, simplifying load balancer configuration and reducing dropped connections during autoscaling events. The trade-off is additional complexity in the caller's state management and latency due to the polling interval or webhook delivery.

8.4 Hybrid and Multi-Tier Serving Architectures

Most production LLM stacks operate more than one model simultaneously. The naive reason is specialization: a small model for classification, a large model for generation. The more powerful reason is cost routing: requests vary widely in the capabilities they actually require, and serving every request on the largest available model wastes the economics of smaller models. Multi-tier serving formalizes this insight into an architectural pattern with explicit routing logic.

In a two-tier architecture, a routing layer sits in front of two or more models and assigns each incoming request to the tier that matches its complexity. Simple queries (FAQ retrieval, intent classification, template filling) route to a smaller, faster, cheaper model. Complex queries (open-ended reasoning, multi-step synthesis, code generation with dependencies) route to a larger model. The routing decision can be made by a lightweight classifier trained on labeled examples from the same workload, by heuristics based on prompt length and token budget, or by a cascade pattern in which the small model attempts the request first and escalates to the large model if its confidence falls below a threshold.

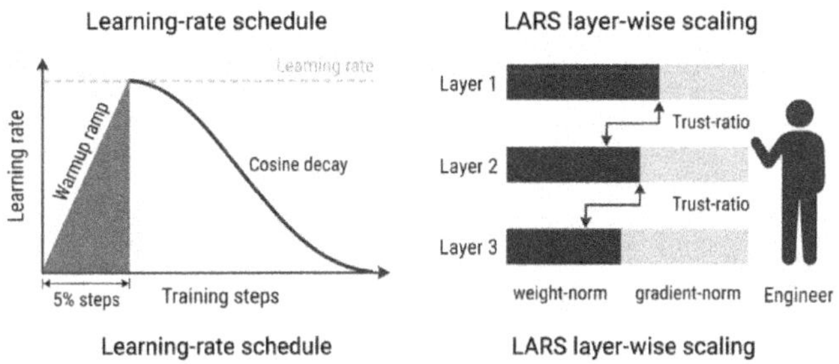

The cascade variant of multi-tier serving is particularly powerful for workloads with heavy-tailed complexity distributions. If 90% of queries are simple enough for the small model to handle at acceptable quality, only 10% of requests incur large-model costs. The challenge is calibrating the escalation threshold. Too aggressive, and the small model handles queries it cannot answer well, degrading output quality. Too conservative, and the routing overhead adds latency to queries that would have been handled adequately without escalation. Empirical calibration on a held-out sample of production queries, scored against the large model's output as a reference, is the most reliable method.

Hybrid cloud-edge architectures extend the multi-tier pattern to a geographic dimension. A small, quantized model runs on edge infrastructure near the user (a CDN node, a regional cluster, or an on-device runtime), handling the vast majority of requests with low latency and without a round-trip to a central data center. Requests that require capabilities beyond the edge model's reach are forwarded to a central large-model deployment. This pattern is particularly valuable for consumer-facing mobile applications and for latency-sensitive workloads in geographic markets where cloud round-trip time is a significant fraction of the TTFT budget.

> *The parallelism patterns covered in this section are presented at the depth an engineer needs to operate a serving stack: how tensor and pipeline parallelism shape memory layout, latency, and throughput once a model no longer fits on a single accelerator, and how those choices interact with batching and KV cache management at run time. This is the operating view, not the design-time architecture of distributed model execution. A reader who needs the architect-view treatment of the same patterns, including how to reason about partitioning strategies, collective-communication topologies, and large-model decomposition before a serving stack exists, will find that material developed at length in the companion volume of this series, Distributed Machine Learning Patterns by Jazper Carter, in its chapters on distributed serving and large-model patterns.*

8.5 Model Sharding, Tensor and Pipeline Parallelism for Inference

When a model's parameter count exceeds the memory capacity of a single accelerator, the model must be distributed across multiple devices. Two principal parallelism strategies govern how this distribution is organized for inference, and each makes a different trade-off between latency and throughput.

Tensor parallelism partitions the weight matrices of each transformer layer across multiple devices. During a forward pass, each device computes its shard of the matrix multiplication, and the partial results are combined via an all-reduce collective before the next layer begins. The benefit is that every device participates in every layer simultaneously, which means the latency of a single forward pass scales with the number of devices (each device does less work per step) rather than multiplying with model depth. The cost is the all-reduce communication across layer boundaries; this communication overhead is tolerable when devices are connected by high-bandwidth interconnects (NVLink or InfiniBand) and becomes prohibitive when they are connected only by PCIe or a network switch.

Pipeline parallelism partitions the model's layers into contiguous groups called pipeline stages and assigns each stage to a different device. A request travels through the stages sequentially, with each device waiting for the previous stage to produce its output before beginning its own computation. The latency of a single request is the sum of the stage latencies and the communication overhead at each stage boundary; pipeline parallelism therefore increases the single-request latency relative to a single-device baseline. What it recovers in return is throughput: if multiple requests are in flight simultaneously, they can be interleaved through the pipeline so that different stages work on different requests at the same time, a technique called microbatching. Pipeline parallelism is appropriate for batch workloads where throughput matters more than per-request latency.

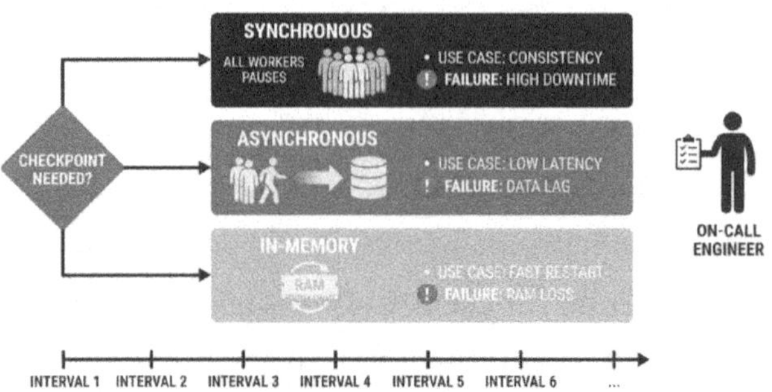

In practice, large-model deployments combine both strategies through a technique called two-dimensional parallelism. A cluster is organized into groups of tightly connected devices (connected by NVLink, for example), and tensor parallelism is applied within each group to minimize communication latency. Pipeline parallelism is then applied across groups, where the cross-group communication is less frequent (only at stage boundaries) and therefore more tolerant of the lower bandwidth of inter-node connections. The degree of tensor parallelism within a node and the number of pipeline stages across nodes are both configuration parameters that must be tuned against the specific model architecture, hardware topology, and target batch size.

A practical calibration note: tensor parallelism beyond eight devices yields diminishing returns for most transformer architectures because all-reduce overhead grows with device count while per-device compute savings shrink. Teams operating large models typically find that a tensor-parallel degree of four to eight devices, combined with pipeline parallelism across nodes, produces the best balance between latency and throughput.

8.6 Quantization, Distillation, and Compression Patterns

The economics of LLM serving are dominated by memory bandwidth. A forward pass through a large transformer spends most of its wall-clock time reading weights from high-bandwidth memory (HBM) and writing KV cache entries, not in arithmetic. Compression techniques that reduce the size of weights and activations directly reduce memory bandwidth pressure, which translates into lower latency and higher throughput at the same hardware cost.

Post-training quantization (PTQ) reduces the numerical precision of a model's weights after training is complete, without any additional training steps. The most common targets are eight-bit integer (INT8) and four-bit integer (INT4) precision. INT8 quantization typically reduces model size by roughly half relative to float16, with negligible quality degradation on most benchmarks; the tradeoff becomes visible only on tasks that are numerically sensitive to weight precision, such as exact arithmetic or low-frequency token prediction. INT4 quantization halves the model size again but introduces more pronounced quality degradation; calibration techniques that identify and preserve the precision of the most numerically sensitive weight clusters (a class of approaches sometimes called mixed-precision or outlier-aware quantization) recover much of the lost quality. PTQ requires a small calibration dataset drawn from the target workload distribution, and its quality impact should be measured on that distribution rather than on generic benchmarks.

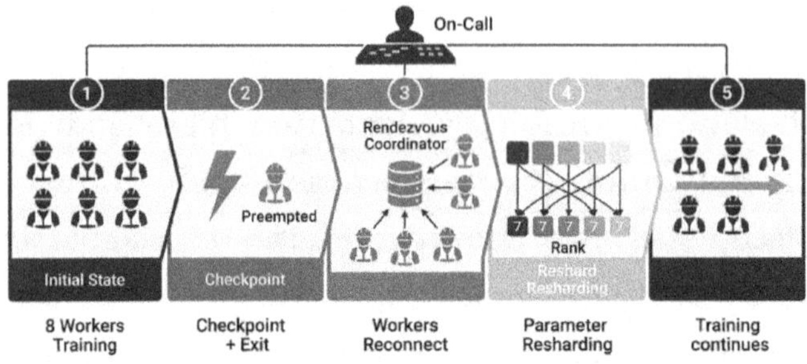

Quantization-aware training (QAT) embeds the quantization operation into the training or fine-tuning loop, allowing the model's weights to adapt to the reduced precision during gradient descent rather than absorbing it as a post-hoc perturbation. QAT consistently produces higher-quality results at lower precision than PTQ, but it requires access to the training infrastructure and sufficient training data to be effective. For teams fine-tuning a base model on proprietary data, QAT is often practical because the fine-tuning infrastructure is already in place. For teams working with a third-party model they do not control, PTQ is the only option without the model provider's cooperation.

Knowledge distillation creates a new, smaller model (the student) by training it to mimic the output distribution of a larger model (the teacher) rather than directly fitting labels. The student model can be dramatically smaller than the teacher: a seven-billion-parameter student can produce output of comparable quality to a thirty-billion-parameter teacher on narrow, well-defined task domains. The cost is the distillation training run itself, which requires a substantial dataset of teacher-generated outputs on the target task distribution, and the quality degradation that occurs outside the training distribution. Distilled models are not general; they are optimized for the task domain covered by the teacher-generated data, and they should not be deployed on out-of-distribution inputs without explicit evaluation.

Speculative decoding is a throughput acceleration pattern rather than a compression technique. Still, it belongs in this section because it is frequently evaluated alongside compression as part of the same cost-reduction effort. In speculative decoding, a small draft model generates a candidate token sequence in a single forward pass, and the large target model verifies the entire candidate sequence in parallel. If the draft tokens match what the target model would have generated (under the target distribution), they are accepted; if they diverge, the sequence is truncated at the first disagreement, and generation continues from the target model. The acceleration depends on the acceptance rate of draft tokens; draft models well-matched to the target model's distribution on the target workload can achieve acceptance rates above eighty percent, translating into throughput gains of two to four times at the same hardware cost.

8.7 KV Cache Management and Paged Attention

The key-value cache is the most consequential memory structure in LLM inference and the one most often treated as an implementation detail rather than infrastructure. Understanding it at an architectural level is a prerequisite to managing latency, cost, and memory pressure correctly at scale.

During inference, each transformer layer computes attention by comparing the current token's query vector against the key and value vectors of every token that preceded it in the context. For autoregressive generation, these key and value vectors are computed once during the prefill phase and then reused in every subsequent decode step without recomputation. This reuse is the KV cache: a per-session store of key and value tensors for every layer and every context token. The memory cost of the KV cache is proportional to the product of the number of layers, the number of attention heads, the head dimension, the context length, and the batch size. For a model with thirty-two layers and a context length of eight thousand tokens serving a batch of sixteen concurrent sessions, the KV cache alone can consume tens of gigabytes of HBM, competing directly with the model weights for the same high-bandwidth memory pool.

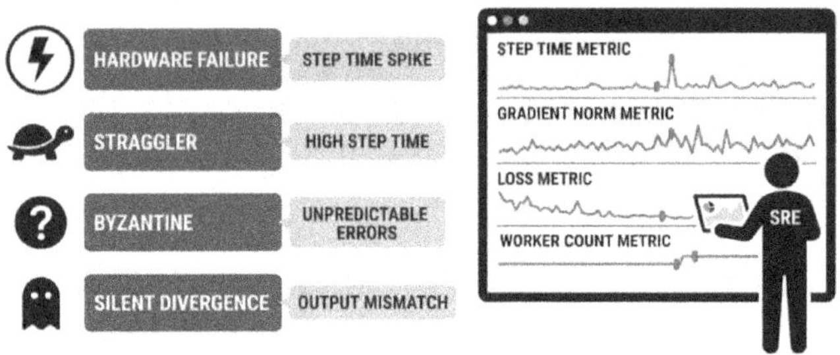

Paged attention addresses the memory fragmentation problem that arises when KV caches are allocated as contiguous memory regions. When sessions have variable context lengths, contiguous allocation leaves gaps of unusable memory between session buffers, and the worst-case memory requirement must be reserved up front even when a session has not yet used its full context budget. Paged attention borrows the concept of virtual memory paging from operating systems: each session's KV cache is divided into fixed-size pages, and pages are allocated from a global pool as they are needed rather than reserving the worst-case budget at session start. A block table maps each session's logical page sequence to physical page addresses in the pool. The immediate benefit is that memory fragmentation is eliminated; pages from completed sessions are returned to the pool and reused immediately by new sessions, increasing the number of concurrent sessions a fixed memory budget can sustain.

Prefix caching extends paged attention with a cross-session sharing mechanism. When multiple sessions begin with an identical prefix (the system prompt is the most common case, but shared few-shot examples and shared document context are also common), the KV cache pages for that shared prefix can be computed once and shared across all sessions that use it. The pages are reference-counted and evicted only when no active session requires them. In practice, a deployment where every session begins with the same one-thousand-token system prompt can eliminate the prefill cost for that prefix after the first session in a time window that computes it, reducing the effective TTFT for all subsequent sessions by the time spent prefilling the shared prefix.

Eviction policies for paged attention pools require careful design. The simplest is least-recently used (LRU): evict pages from the least-recently active session. LRU works well for short-context workloads but degrades for long-context sessions that pause and resume: a session pausing for thirty seconds may have its pages evicted by a burst of new sessions, requiring full prefill on resumption. A priority-weighted policy that accounts for the recomputation cost of evicted pages produces better behavior for mixed-context-length workloads.

8.8 Continuous Batching and Speculative Decoding in Serving

Classical neural network serving treats each request as a discrete job: the server waits until a batch is full or a timeout expires, processes the batch in a single pass, and returns all results before accepting the next batch. This static batching model is deeply inefficient for autoregressive generation because requests in the same batch have different output lengths. A batch of sixteen requests, where one request generates five tokens, and another generates five hundred tokens, keeps fifteen GPUs idle for the duration of the five-hundred-token sequence after the short request completes. GPU utilization, measured as the time actually spent on useful arithmetic, can fall below 30% under static batching with heterogeneous output lengths.

Continuous batching (also called iteration-level scheduling) solves this problem by treating each decoding step as an independent scheduling opportunity. After every token generation step, the scheduler examines the set of in-flight sequences and the incoming request queue. Completed sequences are removed from the batch immediately, and their KV cache pages are returned to the pool. New requests are inserted into the batch to fill the vacated slots, beginning their prefill phase while the remaining sequences continue their decode phase. The batch composition changes at each step, and the GPU is continuously occupied with useful work rather than waiting for the slowest sequence to complete.

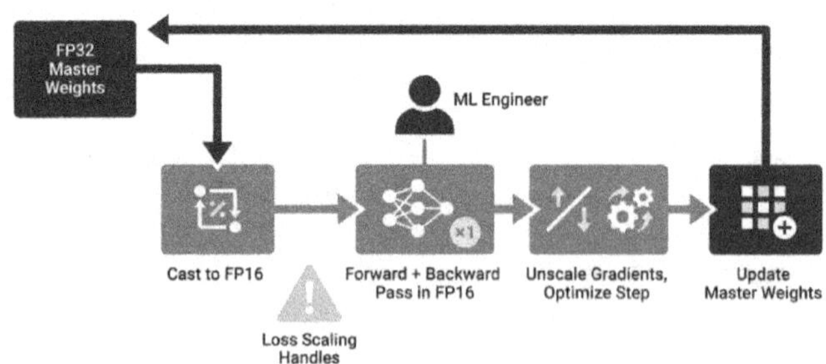

Continuous batching changes the primary resource constraint from batch wait time to KV cache capacity: the system can sustain high utilization only as long as there is available memory to allocate new session pages. When the KV cache pool is exhausted, new requests must wait in the gateway queue rather than being added to the active batch. The practical implication is that KV cache size should be treated as the primary capacity parameter for a continuously batched serving system, not the number of accelerators or the peak request rate in isolation.

Lookahead decoding is a variant that attempts to generate multiple candidate tokens per decoding step without a separate draft model. It maintains a lookahead buffer of candidate token sequences sampled from a Jacobi iteration on the model's own output distribution, then verifies candidates in parallel during the regular decode step. The acceptance rate is lower than that of speculative decoding with a well-tuned draft model. Still, lookahead decoding requires no additional model or memory beyond the lookahead buffer, making it applicable in deployments where the overhead of maintaining a separate draft model is not justified.

Speculative decoding requires that the draft and target models share a compatible tokenizer and vocabulary; otherwise, candidate sequences cannot be verified. The draft model's cost must be accounted for: it runs on every decode step, so its parameter count should be small enough that its compute contribution is well below the target model's savings from reduced steps. Empirically, a draft model with roughly 10% of the target model's parameter count tends to offer a favorable cost-benefit trade-off when the acceptance rate exceeds 70% under the target workload distribution.

8.9 Autoscaling Strategies for LLM Workloads

Autoscaling an LLM serving cluster poses challenges that generic HTTP-service autoscalers are not designed to handle. The fundamental mismatch is that generic autoscalers measure proxies for resource pressure (CPU utilization, request rate, memory usage at the VM level) rather than the actual binding constraint for LLM serving: accelerator memory allocation. A cluster can have abundant CPU headroom and moderate request rates, while every accelerator is at 98% HBM utilization, and new requests are queued. A scaler that reads only CPU or aggregate request rate will not trigger scale-out until those thresholds are crossed, which may happen long after the accelerators are already saturated.

Accelerator-aware autoscaling uses KV cache utilization and request queue depth as the primary scaling signals. When the fraction of KV cache pages in use across the cluster exceeds a configured high-water mark (typically seventy to eighty percent), the scaler adds capacity. When it drops below a low-water mark, the scaler removes capacity after a cooldown period long enough to ensure that in-flight sessions can complete without being disrupted by node removal. Queue depth at the gateway is a leading indicator: requests begin to queue before KV cache utilization reaches the high-water mark, signaling that the scaler can begin provisioning new nodes before existing nodes are fully saturated.

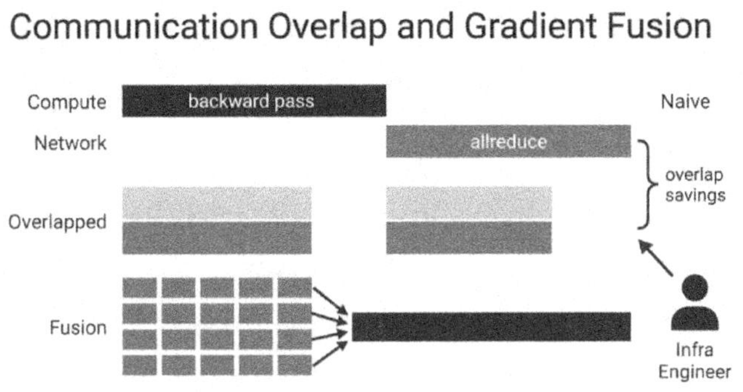

Scale-in for accelerator-backed serving requires special attention because accelerators are expensive and provisioning new ones takes much longer than provisioning CPU-based compute. Aggressive scale-in during a traffic lull followed by a traffic spike leaves the system unable to respond until new accelerators are provisioned, a window that can span several minutes on cloud platforms. Predictive scaling, which uses historical traffic patterns to pre-provision capacity before anticipated demand (before a scheduled marketing event or the start of a business day), is a practical mitigation. The cost of over-provisioning for a predictable demand window is almost always less than the cost of the user experience degradation during the provisioning lag.

Connections between serving instances and external components must be managed carefully during scale-in. A draining protocol stops routing new requests to a node scheduled for removal, waits for in-flight sessions to complete or expire, and terminates the node only then, preventing the dropped-request pattern that characterized the launch-day failure. A maximum drain timeout prevents a single stuck long-context session from indefinitely blocking the removal of an otherwise idle node.

8.10 Caching Layers: Result, Semantic, Prefix, and Embedding

The four canonical cache layers in a production LLM stack operate at different granularities and serve different cost-reduction goals. Deploying all four is not always appropriate; each has trade-offs in consistency that must be weighed against the workload's freshness requirements. Understanding the precise role of each layer, however, is a prerequisite to making an informed decision about which to adopt.

Result caching stores the complete output of a model call keyed on the exact input. It is the highest-leverage cache for workloads with a high proportion of repeated identical queries. A customer support assistant that receives the same ten frequently asked questions from thousands of users per day will serve the vast majority of its traffic from a result cache after the first occurrence of each question, with no model invocation at all. The consistency requirement is that the cached result must remain valid; for queries whose correct answer can change (price lookups, inventory checks, policy questions), result caches must include an expiry time calibrated to the rate of change of the underlying information. Result caches with stale data are worse than no cache at all: they serve incorrect answers with confidence at high speed.

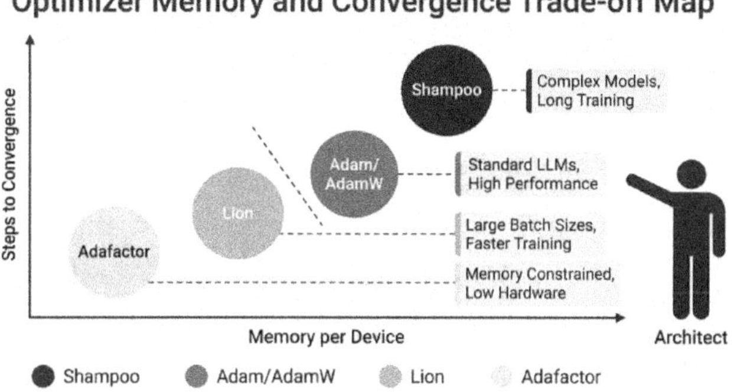

Semantic caching extends result caching by matching queries based on meaning rather than exact string equality. A vector embedding of the incoming query is compared against embeddings of previously cached queries, and if the nearest neighbor exceeds a similarity threshold, the cached result is returned. Semantic caching is particularly valuable for natural language interfaces where users rephrase the same underlying question in many different forms. The operational challenge is setting the similarity threshold correctly: too low, and semantically distinct queries retrieve incorrect cached results; too high, and the cache hit rate is negligible. The threshold should be calibrated empirically on the workload's actual query distribution, and result-quality spot-checking at the threshold boundary should be a standard practice in monitoring.

Prefix caching, discussed in the KV cache section above, operates at the inference-server level rather than the gateway level. It caches the model's internal KV state for shared input prefixes across sessions, eliminating redundant prefill computation. From a caching architecture perspective, prefix caching complements result caching: result caching eliminates model invocations for matching queries; prefix caching reduces the cost of model invocations that do proceed by reusing the KV state for their shared prefix.

Embedding caches store the dense vector representations of input text that are used in retrieval-augmented generation, semantic search, and semantic routing pipelines. Generating an embedding for a document chunk involves a model call that, while much cheaper than a generation call, still incurs compute and latency costs. For document corpora that change infrequently, embedding caches that store pre-computed vectors indexed by document chunk identifier can eliminate embedding computation for the vast majority of retrieval calls. The invalidation policy should be tied to the document's modification time. When a document is updated, its cached embeddings are invalidated and recomputed on next access or asynchronously on a background refresh schedule.

8.11 Latency Budgets, TTFT, and Streaming UX Patterns

Latency in LLM serving is not a single number. It is a profile composed of at least three distinct measurements: time-to-first-token (TTFT), the inter-token latency (ITL) or tokens-per-second (TPS) rate during generation, and the total response time from request submission to the last token delivered. Each of these measurements answers a different question and is appropriate for a different audience. TTFT answers the question a user experiences directly: how long before the system shows any sign of responsiveness? Inter-token latency governs whether the streaming output feels smooth or jerky. Total response time matters for downstream systems that consume the output programmatically and need to bound their own timeouts.

Setting latency budgets without distinguishing these three measurements leads to the failure mode from the opening scenario: a system that meets its average total-response-time target while exhibiting TTFT at the ninety-fifth percentile that is unacceptable to users. Latency budgets should specify all three measurements at the 50th and 95th percentiles. A representative budget for a real-time conversational assistant might set TTFT at less than one second at the fiftieth percentile and less than two seconds at the ninety-fifth percentile, TPS at more than twenty tokens per second, and total response time as unconstrained (the user is reading as tokens arrive). For a programmatic API caller, the budget might be total response time under ten seconds at the ninety-fifth percentile, with TTFT less important because the caller is not presenting a streaming interface to a human user.

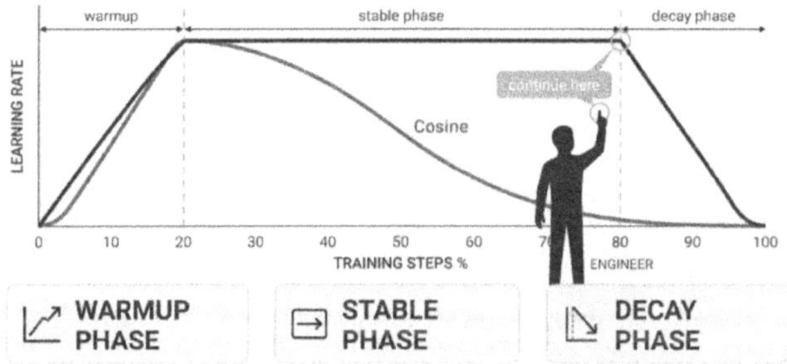

Streaming user experience patterns reduce perceived latency below measured latency by giving the user something meaningful to read before the generation is complete. The most effective techniques are progressive disclosure (beginning the response with the most useful information, so that even a partial output is actionable), skeleton screens (showing placeholder tokens or typing indicators during prefill so the user understands that work is in progress), and chunked streaming (flushing tokens to the client in small groups rather than one at a time to reduce rendering overhead and perceived stuttering). These are presentation-layer concerns, but they are worth engineering attention: a system that measures one-second TTFT and presents a skeleton screen during prefill is perceived as substantially faster than a system that measures the same TTFT but presents a blank screen until the first token arrives.

Gateway-level streaming management requires care at scale. Each streaming connection holds server resources for the full generation duration; at high concurrency, the number of open connections can exceed the default limits of most load balancers and API gateways. Connection limits and timeouts should be configured explicitly for streaming workloads rather than inherited from defaults tuned for short REST requests. A timeout measured from connection open rather than from the last token delivery will silently truncate long outputs; the correct policy tracks inactivity between token deliveries.

8.12 Edge, Browser, and Mobile Serving Patterns

The dominant mental model for LLM serving assumes a centralized cloud deployment, but a growing class of workloads benefits from pushing inference closer to the user. On-device and edge-serving patterns reduce round-trip latency, eliminate the need for a network connection during inference, preserve privacy by keeping input data off remote servers, and reduce cloud costs per inference. The constraints are the computational and memory limits of the edge device, which currently make on-device deployment practical only for small quantized models or specific-purpose adapters on top of a small base model.

Browser-based inference using WebAssembly and WebGPU runtimes has matured to the point where models in the one-to-three-billion parameter range, quantized to four-bit precision, can run in a modern browser without any installation or server dependency. The initial download of model weights (typically one to three gigabytes) is the primary friction; once cached by the browser, subsequent sessions start inference immediately. This pattern is appropriate for use cases where privacy is paramount (on-device transcription, personal note summarization, offline document assistance) and where the user population can be expected to have devices capable of the workload. It is not appropriate for workloads requiring models larger than can fit in browser memory, or for workloads where output quality from a small quantized model is insufficient.

PEFT at Scale: LoRA Distributed Fine-Tuning Architecture

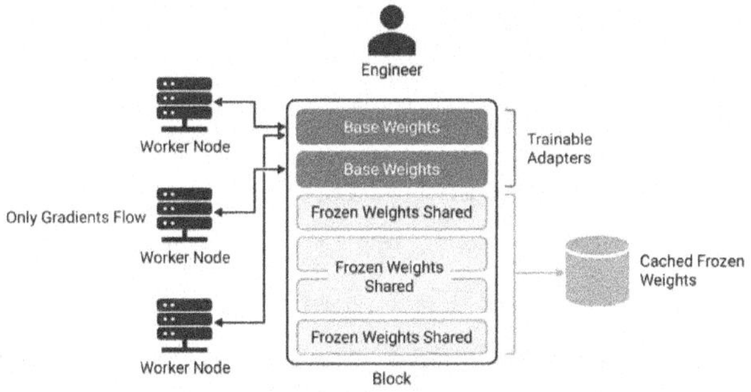

Mobile native inference using neural engine frameworks on modern mobile chips extends on-device serving beyond browser runtimes. Dedicated neural processing units on current mobile hardware sustain twenty to forty tokens per second for models in the one-to-four-billion parameter range, adequate for conversational features. The model must be compiled to the device's native accelerator format and distributed within the application bundle or downloaded on first use. The engineering cost is integrating the inference runtime into the mobile build pipeline and managing model versioning across an installed user base, where updates cannot be forced.

Hybrid edge-cloud topologies route requests between edge and central deployments based on complexity, sensitivity, and network availability. A query the edge model handles with high confidence stays local; one requiring broader knowledge or longer context forwards to the central cluster. Routing uses a lightweight classifier or the edge model's own confidence score. Network unavailability triggers graceful degradation to edge-only operation with an explicit quality caveat. This degradation path is often omitted from the first mobile LLM feature version and discovered only when users in low-connectivity environments report failures.

8.13 Safe-Deployment Patterns: Canary, Shadow, Blue-Green, and Rollback

Model updates in an LLM serving system are qualitatively riskier than code deployments in classical services. Automated tests can often validate code deployments before they reach production traffic. A model update changes the output distribution in ways that automated tests may not capture; a model that passes a benchmark suite may still produce outputs on tail production inputs that violate safety requirements, exceed cost budgets, or degrade user experience in ways that are only visible through real traffic. Safe-deployment patterns bound the blast radius of any single model change and create a structured pathway for detecting and remediating problems before they affect the full user population.

Canary deployment routes a small fraction of production traffic (typically one to five percent) to the new model version while the remainder continues to be served by the current version. Quality signals (user engagement metrics, safety classifier pass rate, and latency at the canary tier) are monitored during the canary period.

If the signals are acceptable, traffic is progressively shifted to the new version. If a signal degrades, the canary is stopped, and traffic is shifted back entirely. The critical design decision is which signals constitute a gate: latency and error rate are straightforward, but output-quality signals require explicit investment. A canary that monitors only latency and error rate will not catch a model that generates subtly lower-quality or subtly higher-risk outputs than its predecessor.

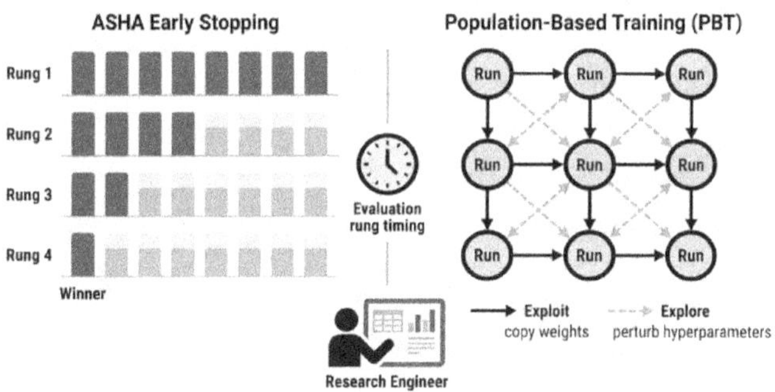

Shadow deployment runs the new model version in parallel with the current version on a copy of production traffic without exposing its outputs to users. The shadow model's outputs are compared offline against the current model's outputs for the same inputs, identifying regressions in quality, safety, and latency before the new version ever reaches users. Shadow deployment is the highest-confidence pre-production validation available for model changes because it uses real production inputs rather than a held-out test set. The cost is the computational overhead of running two models simultaneously; this is typically acceptable for a brief validation period before a major version change and can be reduced by shadowing only a sampled fraction of traffic.

Blue-green deployment maintains two full-capacity environments, the current (blue) and the new (green), and switches traffic between them atomically via a load balancer reconfiguration. In the classical software deployment context, the switch is instantaneous, and rollback is equally instantaneous. For LLM serving, the switch must account for in-flight streaming sessions: a session that began on the blue environment holds a connection to that environment's serving instance and its KV cache state. An atomic switch that terminates all blue connections would drop every active session. The practical mitigation is a drain period during which new connections are routed to the green environment while existing blue connections are allowed to complete or time out.

Rollback must be a first-class operational capability, not an afterthought. Every model deployment should be accompanied by a documented rollback plan that specifies the rollback trigger conditions (which signal at which thresholds warrant an immediate rollback versus a graduated one), the rollback procedure (load balancer configuration change, model artifact swap, or environment switch), and the expected time to rollback from the trigger decision. Regularly practice rollbacks in non-production environments; a rollback procedure that has never been executed will fail under the pressure of a live incident. The model artifact for the current production version should always be retained and immediately available, not archived or deleted after a successful deployment.

8.14 Technical Checklist

Use the following checklist as a release gate before declaring a serving system production-ready against the patterns in this chapter.

1. Latency is instrumented at the percentiles that matter: TTFT at p50 and p95, inter-token latency, and total response time are all measured and alarmed on separately.
2. The serving topology (synchronous, streaming, batch, or hybrid) was chosen deliberately against the workload's latency and throughput requirements, not inherited from a tutorial or default configuration.
3. Compression options (INT8 PTQ, INT4 PTQ, QAT, distillation, speculative decoding) have each been evaluated against the target workload's quality requirements and cost constraints, and the chosen option is documented with its measured quality impact.
4. KV cache utilization is monitored as a first-class infrastructure metric; alerts exist for high-water-mark breaches that predict incoming request queuing.
5. Paged attention or an equivalent memory management strategy is in place to prevent KV cache fragmentation under variable-context-length workloads.
6. Prefix caching has been evaluated; if the workload has a shared system prompt or shared document context, prefix caching is active, and its hit rate is measured.
7. Result caching and semantic caching have been evaluated; if adopted, their hit rates, latency reduction, and cache invalidation policies are measured and reviewed quarterly.
8. Continuous batching, or an equivalent iteration-level scheduling strategy, is in place for interactive workloads to prevent GPU idle time caused by heterogeneous output lengths.
9. The autoscaler uses accelerator-aware signals (KV cache utilization and request queue depth) rather than relying solely on CPU or aggregate request rate.

10. Scale-in uses a drain protocol that completes or times out in-flight sessions before terminating nodes, preventing session drops during scale events.
11. Streaming UX patterns (skeleton screens, chunked delivery, progressive disclosure) are in place to reduce perceived latency for user-facing features.
12. Gateway streaming configuration (connection limits and timeouts measured from the last token delivery rather than connection open) has been explicitly tuned for the streaming workload.
13. Canary deployment is a structural requirement for every model update; quality signal gates are defined before the canary begins, not after a problem is detected.
14. Shadow deployment has been used for at least the most recent major model version change; offline comparison results are retained.
15. Blue-green deployment includes a connection drain protocol for in-flight streaming sessions.
16. A rollback plan with documented trigger conditions and a measured rollback time is on file for every production model version.

8.15 Team Conversation

The following questions are intended for technical leads to use in team meetings to pressure-test alignment on serving architecture and latency engineering. There are no single correct answers; the value is in surfacing assumptions and gaps.

1. Do our latency dashboards include TTFT at the ninety-fifth percentile, or only average total response time? If not, what would we need to instrument to add it?

2. Where in our workload portfolio would a multi-tier routing architecture provide the largest cost reduction? What would we need to measure to quantify the opportunity?
3. Which compression technique would unlock the most savings for our most expensive serving workload? Have we measured the quality impact of each option on our specific task distribution, not just on published benchmarks?
4. Where in our stack is the KV cache treated as a side effect of the inference library rather than as a managed infrastructure resource? What would it mean to treat it as a first-class resource?
5. Which of the four cache layers (result, semantic, prefix, embedding) would have the largest latency and cost impact if we added it in the next quarter? What would we need to validate before adopting it?
6. Have our autoscalers ever made a decision we could not explain after the fact? What signals would we need to add to make every scaling decision auditable?
7. When was our last canary deployment, and what quality signal gates did we monitor? If we removed the quality gates and monitored only latency and error rate, what regressions might we have missed?
8. Have we ever practiced a rollback in a non-production environment? If a production rollback were triggered tonight, what is the step-by-step procedure, and who owns it?

8.16 Key Takeaway

Serving a large language model in production is its own engineering discipline, with primitives (token streaming, KV cache allocation, prefill/decode phase separation, paged memory management) that do not exist in classical model serving. Teams that approach LLM serving as a straightforward extension of their existing API infrastructure will repeatedly encounter the same cluster of failures: tail latencies that are multiples of average latency, autoscalers that thrash on the wrong signals, and cost profiles that scale linearly with traffic, even when caching and batching could make them sublinear. The chapter's patterns are not academic; they are the specific interventions that distinguish serving stacks that remain economical and reliable at scale from those that require increasingly expensive hardware to paper over architectural gaps.

Decisions, not hardware, shape the throughput-versus-latency curve in LLM serving. Continuous batching changes GPU utilization from 30% to 85% on the same accelerators. Prefix caching eliminates prefill cost for the most common queries. Semantic caching reduces the fraction of requests that reach the model. Speculative decoding increases effective token throughput by a factor of two to four. Each intervention operates on the architecture rather than on hardware, and each is applicable before any additional hardware spend. An engineer who understands the full pattern catalog can squeeze substantially more throughput and lower latency from a fixed budget than one who treats serving as a hardware procurement problem.

Safe-deployment patterns are not operational polish to be added after the serving architecture is stable; they are structural requirements that determine the risk surface of every model change. A serving system that cannot canary or roll back a model update absorbs the full blast radius of every deployment into production traffic. The patterns in this chapter represent the field's accumulated understanding of how to make LLMs serve reliably, economically, and safely in practice rather than by luck. Applying them systematically, instrumenting the signals that confirm they are working, and reviewing them quarterly against the actual workload distribution is how a team moves from a pilot that works in demos to a production system that earns and sustains the confidence of the organization that depends on it.

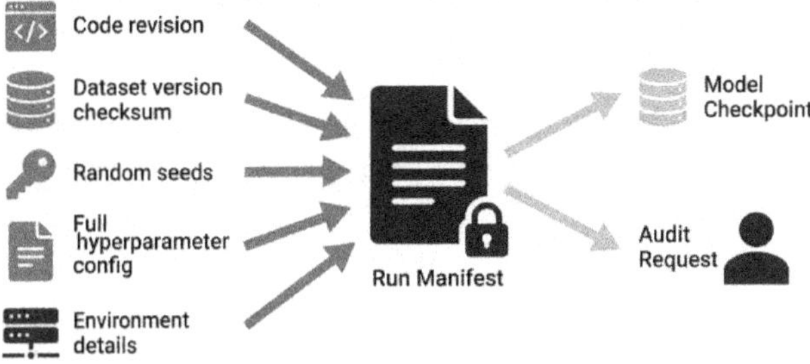

9 Cost Engineering and Resource Management

9.1 Opening Scenario

The alert came in mid-morning on a Thursday: the cloud billing forecast had crossed a threshold that triggered a Slack notification no one had configured intentionally. A senior platform engineer pulled up the monthly dashboard and found that LLM inference costs had surpassed every other infrastructure category combined, including compute, storage, networking, and the data warehouse that had historically dominated the bill. The spike was not gradual. It had doubled in eleven days.

A spot audit across the platform's three consumer teams revealed a pattern that was both unsurprising and costly. One team had recently migrated a document-summarization workflow to a frontier-tier model because the latency of the previous model was slightly higher than desired. The documents averaged two hundred words. A mid-tier model would have handled them the same way. The cost difference was a factor of twelve per token. The team had not documented the change or flagged it to the platform group, because no approval process existed for model selection.

A second team had a nightly batch job that chunked customer feedback into sentiment categories. The job ran against the same interactive inference endpoint used by real-time product features. Batch and interactive traffic shared a pricing tier designed for low-latency, synchronous workloads. The batch job was paying interactive prices for work that had no latency requirement and could have been processed against a dedicated asynchronous endpoint at roughly a third of the cost.

A third team had enabled streaming output for a pipeline that generated structured JSON payloads—streaming added significant overhead to the response lifecycle and prevented the serving layer from applying response-level caching. The team had enabled it because a developer had read that streaming improved perceived latency for chat interfaces. This was not a chat interface. The effect was a tripled unit cost with no quality benefit.

What unified all three situations was the same absence: no cost owner, no unit-economics target, no dashboard that surfaced per-request cost, and no review gate that would have caught any of these decisions before they reached production. The fix required no new infrastructure. It required a framework: a documented cost model for each workload, a routing policy that matched work to the cheapest capable model, a chargeback signal that made team-level consumption visible, and a tactic catalog that the group could work through each quarter. That framework is what this chapter provides.

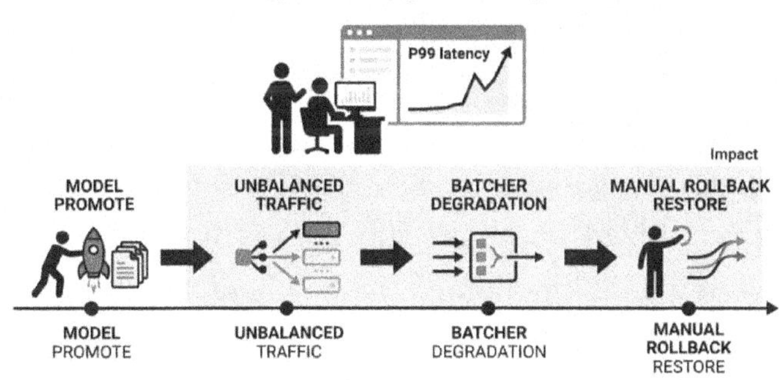

9.2 Why It Matters

LLM workloads break the cost assumptions that engineering teams built their intuitions against. In a conventional web service, a single bad configuration decision might move latency or error rates, but the fixed capacity of the instance fleet bounds the cost impact. In an LLM system, a single prompt change can double the token count per request; a model upgrade can multiply the per-request cost by an order of magnitude; and a runaway agentic loop can exhaust a quarter's budget before the on-call engineer reads the alert. The cost surface is proportional to consumption, and consumption is shaped by dozens of decisions that engineers make every day without a cost signal in front of them.

The workflow-level impact of this dynamic is concrete. Teams that do not instrument cost per request cannot detect regressions before they appear on the monthly bill. Teams that do not own a cost model cannot distinguish a legitimate cost increase driven by growth from an illegitimate one driven by a misconfiguration. Teams that share a single billing account across multiple product features cannot tell which feature is responsible for a spike, which means they cannot fix it quickly or assign accountability clearly. These are not edge cases. They are the default state for most LLM platforms during their first 18 months of production.

Cost engineering treats these dynamics as an engineering discipline rather than a finance concern. It gives the team the vocabulary to set targets, the observability to detect deviations in real time, and the tactic catalog to close the gap between current spend and the optimized baseline. The goal is not minimizing cost; it is achieving mission-aligned cost, where the team can demonstrate that every dollar of LLM spend is producing a proportional unit of business value and that no dollar is being burned on a misconfiguration, an unnecessary model upgrade, or a workload pattern that was chosen by accident rather than design.

9.3 Cost Models for Inference, Fine-Tuning, and Training

The first step in treating cost as an engineering discipline is making the cost model explicit. LLM workloads generate cost through three structurally distinct mechanisms: inference, fine-tuning, and pretraining. Each has a different cost structure, amortization horizon, and set of levers to reduce it. Conflating them produces a bill that appears to be a single number yet resists diagnosis.

Inference cost is the most immediate and the most variable. For hosted API providers, inference is typically priced per token, with separate rates for input and output tokens. Output tokens cost more because they require sequential generation rather than parallel attention computation. The practical implication is that a prompt design that produces shorter outputs without sacrificing quality is a cost optimization, not just a quality decision. For self-hosted models, inference cost is expressed as an infrastructure cost per unit time, typically measured in GPU-hours on the cloud provider's pricing schedule. The relevant metric is throughput: how many tokens per second can the serving stack process per dollar of infrastructure, and how does that ratio change as request volume scales. Self-hosted inference shifts the cost structure from variable per-token charges to fixed infrastructure charges with a variable throughput numerator.

Fine-tuning cost is a training-time cost that recurs each time the model is adapted to new data or a new task. For hosted fine-tuning services, the cost is typically per training token processed. For self-hosted fine-tuning, the cost is GPU-hours required to run the adaptation process. Both figures need to be amortized across the expected volume of inference requests that will use the fine-tuned model before it is retrained or retired. A fine-tune that costs a thousand dollars to produce is economically different depending on whether the resulting model serves 10,000 requests or 10 million. Documenting the expected amortization horizon at the time of fine-tuning is the first step in making fine-tuning cost auditable rather than implicit.

Pretraining costs are typically incurred only by organizations building foundation models from scratch, and they are measured in millions of dollars of GPU-cluster time. For the majority of production LLM teams, pretraining is not a direct cost. It matters indirectly because it determines which hosted models exist at which price points and which open-weight models are available for self-hosting. Teams building cost models do not need to account for pretraining costs directly. Still, they should understand that the economics of frontier model providers are shaped by pretraining amortization, which is one reason frontier API prices have a floor unlikely to drop below a certain threshold regardless of competitive pressure.

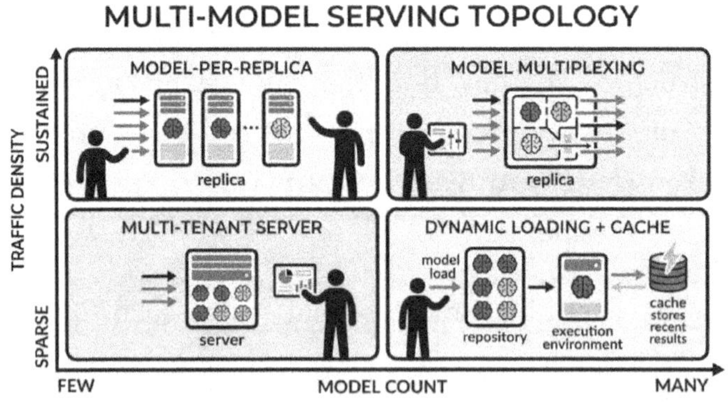

The practical output of building these cost models is a cost register: a document that names each production LLM workload, its cost class (inference, fine-tuning, or both), its billing mechanism (per-token or infrastructure-hour), its current unit cost, and its cost owner. A cost register does not need to be a sophisticated system. A shared document that the platform team reviews monthly is sufficient to catch the pattern seen in the opening scenario, where three workloads were generating avoidable costs for reasons that no one had ever examined together.

9.4 Unit Economics: Cost Per Request, Per User, Per Feature

Cost per month is a management signal. Cost per request is an engineering signal. The distinction matters because the monthly total aggregates every workload, every feature, every experiment, and every misconfiguration into a single number that cannot be acted on directly. Cost per request isolates the unit of work the engineer actually controls: one call to the model, with a specific prompt, returning a specific response. When the cost per request changes, the engineer can identify the specific workload responsible and examine the inputs that drove the change.

The unit-economics hierarchy has three levels, each with a different audience and a different action horizon. At the request level, cost per request is the signal engineers use in real time and in CI/CD gates. A request-level regression that doubles the cost of a specific endpoint is detectable on a rolling window and can be investigated before it compounds. At the session level, the cost per user session aggregates all LLM calls within a bounded interaction, which is the unit that product managers use when they connect engineering spend to product engagement. At the feature or product level, cost per dollar of revenue attributed to the feature connects LLM spend to business outcomes. It is the language that finance understands when asking whether an AI feature is economically viable.

Building the measurement infrastructure for unit economics requires token-level instrumentation at the request boundary. Every inference call needs to record input token count, output token count, the model tier used, the workload identifier, and the team or feature that owns the request. These fields are the basis for all three levels of the hierarchy. Without them, unit economics is estimated rather than measured, which means regressions are discovered on the monthly bill rather than on the real-time dashboard.

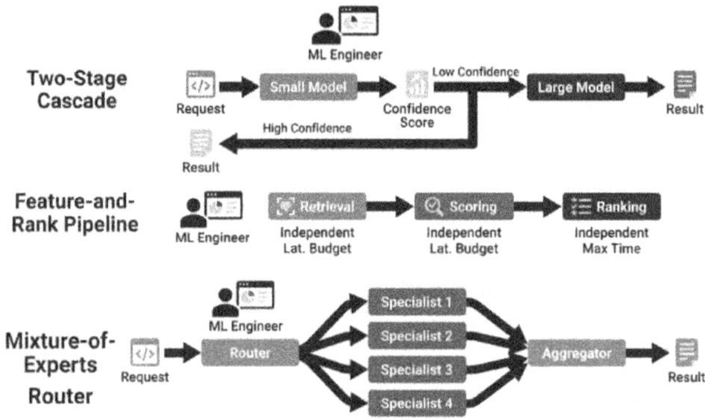

Unit economics work best when set before launch and reviewed at each major traffic milestone. A reasonable initial target for a new feature is derived from the product's margin requirements. If the feature is expected to contribute a specific revenue figure per user session, the cost per session should be a fraction of that contribution, consistent with the product's gross margin targets. Setting the target first and measuring against it operationally is the pattern that turns cost from a surprise into a managed variable.

9.5 Spot Instance Strategies and Fallback Policies

For teams running self-hosted inference, spot and preemptible capacity is the highest-leverage cost lever at the infrastructure layer. Spot instances on major cloud providers are priced at discounts typically ranging from 50 to 80 percent below on-demand rates, with the trade-off that the cloud provider can reclaim the capacity with a short preemption notice (typically two minutes). For workloads that can tolerate interruption and recover gracefully, this discount represents a substantial reduction in the infrastructure component of inference cost.

The operational challenge of spot capacity is designing a serving architecture that absorbs preemption without producing latency spikes or request failures visible to the application layer. The canonical pattern is a multi-pool architecture: a baseline pool of on-demand instances sized to handle the minimum sustained load, supplemented by a larger spot pool that handles incremental traffic above the baseline. When spot instances are preempted, the serving layer sheds the incremental load rather than failing the baseline traffic. The on-demand pool provides the reliability floor; the spot pool provides the cost discount on the variable fraction of traffic.

Fallback policy design is the operational half of spot strategy. A fallback policy specifies what happens when the spot pool shrinks faster than the traffic decreases. The options are ordered by cost and latency impact: first, route overflow to on-demand instances in the same region; second, route to on-demand instances in a different availability zone; third, route to a hosted API endpoint as a temporary backstop, accepting the per-token pricing as a short-duration cost spike to preserve availability. Each fallback tier must be tested under simulated preemption conditions before being trusted in production. A fallback policy that exists only in documentation and has never been exercised is not operational; it is a plan.

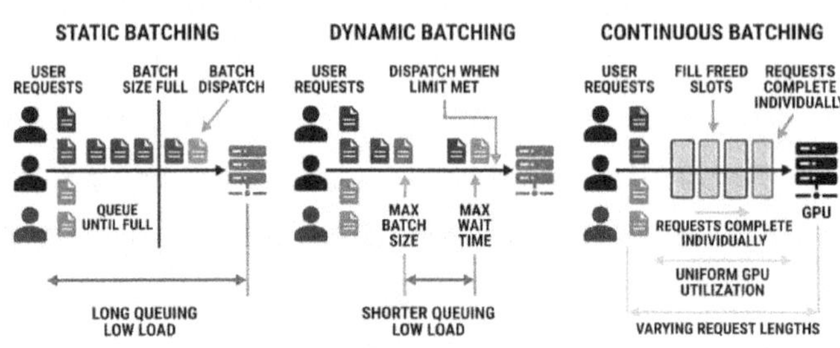

For batch workloads with no latency requirement, a simpler spot strategy is sufficient: run the entire batch job against spot capacity, checkpoint intermediate results periodically so that a preemption only loses the work since the last checkpoint, and requeue interrupted jobs automatically. This pattern can reduce batch inference costs to near the spot discount floor with no reliability impact, since batch jobs have no SLA on completion time that requires the reliability floor of on-demand capacity.

9.6 Request Batching and Shaping Techniques

Batching is the process of grouping multiple inference requests for processing together rather than individually. Its cost-benefit derives from how LLM serving hardware works: a GPU cluster performing inference can often process a batch of requests with nearly the same latency as a single request, because the attention computation over a batch is more efficient than running equivalent sequential computations. When individual requests are small, the hardware is underutilized for every single request. Batching increases utilization, which increases throughput per unit of infrastructure cost.

Server-side continuous batching is the most impactful form and is now a standard feature of production inference serving frameworks. The serving engine maintains a queue of incoming requests. It continuously assembles them into batches as GPU capacity becomes available, rather than waiting for a batch to fill before starting processing. The effect is that the engine achieves high hardware utilization without adding the tail latency of a fixed wait window. For teams running self-hosted inference, enabling continuous batching is among the first optimizations to implement. For teams using hosted APIs, batching behavior is controlled by the provider, but structuring workloads to enable batching (for instance, by avoiding request patterns that require immediate sequential responses) improves the provider's ability to offer favorable pricing on batch-specific endpoints.

Client-side bundling applies to workloads where the calling application controls the request rate and can defer non-urgent requests to accumulate a batch. A document processing pipeline that receives documents in real time does not need to send each document to the model as it arrives. It can accumulate documents for a short window and send them as a batch request, taking advantage of batch pricing tiers offered by hosted providers or improving hardware utilization on a self-hosted cluster. The trade-off is added latency equal to the accumulation window. For workloads where documents do not need a response within seconds, this trade-off is favorable.

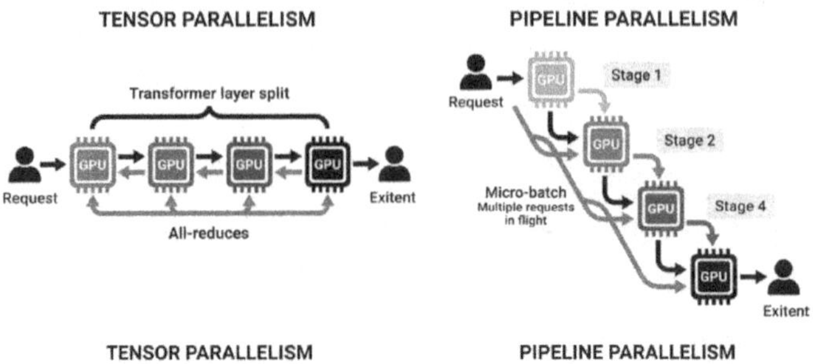

Request-rate shaping addresses the cost impact of traffic spikes. When a burst of requests arrives simultaneously, the serving layer must either queue them (incurring latency) or scale horizontally (incurring additional capacity costs). An admission control layer that smooths the request rate into a sustainable queue reduces the peak capacity required, thereby lowering infrastructure costs for handling bursty traffic patterns. Rate shaping is particularly useful for workloads driven by scheduled events (batch jobs, scheduled reports) or user behavior patterns (morning usage spikes) that are predictable enough to shape in advance.

9.7 Model-Tiering and Cascade Patterns

Not all inference tasks require the same model capability. A request that asks a model to extract a date from a sentence does not require the same parameter count as one that asks it to synthesize a technical argument from fifty documents. When a large model handles both tasks, the cost of the first request is inflated by the capability required only for the second. Model-tiering is the pattern that prevents this by defining a set of model tiers (typically small, medium, and large) and routing each request to the cheapest tier capable of handling it at the required quality level.

The cascade pattern is the dynamic form of tiering. Rather than routing requests to a specific tier at design time, the cascade routes requests first to a small model, evaluates the quality of its response against a threshold, and escalates to a larger model only when the small model's response does not meet the threshold. The threshold can be based on a confidence score from the model itself, on a lightweight classifier trained to distinguish adequate from inadequate responses, or on explicit routing rules derived from the request structure. The cascade's cost savings are realized for the fraction of requests that the small model handles adequately, which is often a large majority in production workloads, where simple requests outnumber complex ones.

Designing a cascade requires three decisions: the threshold definition (what counts as an adequate response from the small model), the escalation path (which larger model receives the escalated request), and the budget policy (whether to escalate unconditionally when the threshold is not met, or to return the small model's best response if the cost budget for the request has been exceeded). The threshold definition is the hardest because it requires an offline evaluation dataset that covers the distribution of real requests and is large enough to set the threshold without biasing toward either too many escalations (which erodes the cost saving) or too few (which degrades quality).

Accelerator-Aware Load Balancing with Power-of-Two Choices

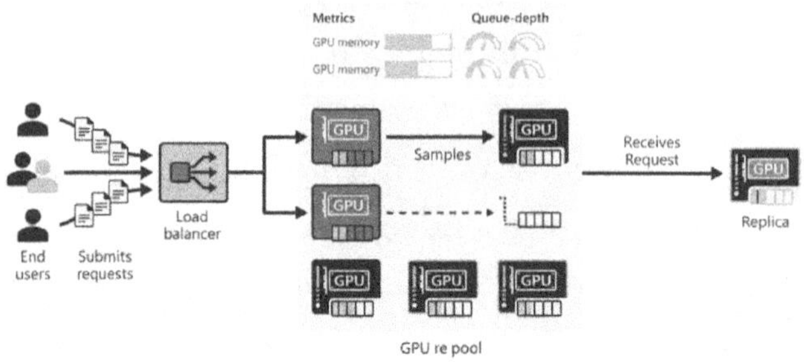

Model-tiering also applies to fine-tuned variants. A team that has produced a fine-tuned small model for a specific task (such as extracting structured data from a specific document format) may find that the fine-tuned small model handles most of those requests better than a generic large model at a fraction of the cost. The fine-tuning investment is justified when the volume of requests for the target task type is sufficiently high that the per-request cost savings exceed the amortized fine-tuning cost over a reasonable horizon.

9.8 Caching as a Cost Strategy

Caching is among the highest-leverage cost interventions available at the application layer. A cached response costs zero tokens because the inference computation is not performed. Its cost impact is therefore proportional to the cache hit rate multiplied by the average cost of the requests that hit the cache. In workloads with high prompt similarity (FAQ-style assistants, templated report generators, search-adjacent summarization), cache hit rates above 50% are achievable, meaning half the requests incur no model cost whatsoever.

The serving architecture typically supports two cache layers with different cost profiles. Exact-match caching stores the complete response for a specific prompt string and returns it on an exact match. It is cheap to implement, trivial to validate (the cached response is identical to the generated one), and effective for workloads where many users send identical or near-identical prompts. Semantic caching stores responses indexed by a dense embedding of the prompt and retrieves the response for the semantically nearest cached prompt within a similarity threshold. It covers the broader class of prompts that differ in wording but share meaning, at the cost of an embedding computation on each request and a retrieval operation against the semantic cache index.

The cost accounting for caching needs to include the operational cost of the cache infrastructure itself. An exact-match cache backed by a fast key-value store has negligible infrastructure cost relative to the inference cost it displaces. A semantic cache backed by a vector index incurs a non-trivial infrastructure cost that must be weighed against the inference cost saved. The break-even point depends on the cache hit rate and the model tier: a semantic cache that reduces inference calls to a large frontier model by thirty percent justifies substantially more infrastructure cost than one that reduces calls to a small self-hosted model by the same percentage.

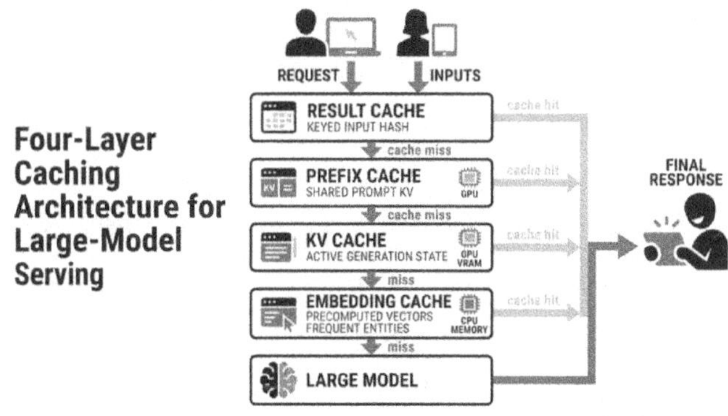

Cache invalidation is the risk that requires explicit engineering. A cached response that was correct when generated may become incorrect if the underlying information the model was reasoning over has changed. For workloads where responses depend on time-sensitive information (current prices, recent events, live data), caching requires a TTL policy that expires cached entries before they become misleading. For workloads where responses are based on stable information (product documentation, legal text, fixed reference material), longer TTLs are safe, and the cost savings are correspondingly larger. Documenting the TTL policy per workload is part of making caching auditable rather than invisible.

9.9 Monitoring Cost Per Request and Real-Time Cost Anomaly Detection

Cost observability transforms the monthly billing review into a continuous signal. The core instrumentation requirement is straightforward: every inference call emits a cost event that records the workload identifier, the model tier, the input and output token counts, and the timestamp. These events flow into the same observability pipeline as latency and error-rate metrics, where they are aggregated into per-workload cost rates that can be visualized on a dashboard and alerted on in real time.

The dashboard design for cost observability mirrors that for latency observability. The primary panel shows cost per request as a rolling time series for each workload, with a baseline band derived from the prior rolling average and alert thresholds set to multiples of the standard deviation. A secondary panel shows cost by model tier, revealing the pattern in which a workload has migrated to a more expensive tier without a corresponding change to the workload identifier. A tertiary panel shows total spend rate by team or feature, which is the signal the platform team uses to identify which team is responsible for a cost spike before investigating at the workload level.

Anomaly detection on cost signals differs from anomaly detection on latency in one important respect: cost anomalies are typically slow-moving compared to latency spikes. A latency spike is visible within seconds. A cost regression that doubles the per-request cost may take minutes to hours to produce a visible signal on the daily spend rate, depending on request volume. The implication is that cost anomaly detection benefits from both short-window alerting (catching runaway agents or accidental infinite loops quickly) and longer-window trend detection (catching gradual model-tier creep or prompt expansion that increases token counts over weeks).

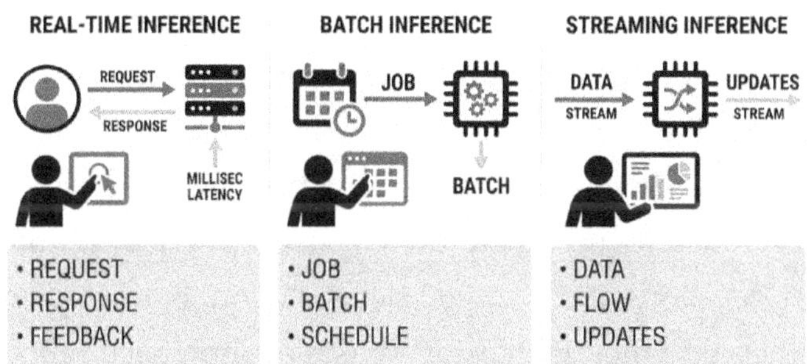

Runaway agents are the highest-priority cost anomaly pattern in systems with agentic workloads. An agent that enters a loop or issues unintended recursive tool calls can generate thousands of inference requests in seconds, each billed at normal per-token rates. Short-window alerting on the total request rate per agent session, combined with a hard token budget enforced by the agent runtime per session, is the standard operational control. The budget acts as a circuit breaker: when the session's token consumption crosses the budget threshold, the runtime terminates the session. It logs the event, preventing an uncapped runaway from propagating to the billing layer.

9.10 Chargeback and Showback Models for Multi-Team Environments

In a platform that serves multiple product teams from a shared LLM infrastructure, a single aggregated cloud bill obscures the consumption pattern of any individual team. Showback and chargeback are the two cost-attribution patterns that restore visibility at the team level. They differ in operational consequence: showback reports each team's consumption as an informational signal, while chargeback transfers the cost to the consuming team's budget or allocates it against a shared cost metric that affects the team's performance evaluation.

Showback is the right starting point for most platform teams. It requires no budget transfer mechanism or negotiation over allocation methodology, so it can be implemented as soon as the per-request cost instrumentation is in place. Its effect is behavioral: when teams see their own consumption data, they make different decisions about model tier selection, batching, and caching than they do when the cost is invisible. Showback reports work best when they include a comparison to peers (which team is spending the most per request on equivalent workloads) and a cost-per-request trend (is this team's unit economics improving or deteriorating over time).

Chargeback requires additional design decisions that Showback does not. The first is the allocation unit: is cost allocated per request, per token, per model tier, or as a flat rate per active feature? Per-token allocation is the most granular and the most actionable, but it requires the consuming team to understand token-level cost drivers. Per-feature allocation is simpler but creates incentives to add features rather than optimize existing ones. The second decision is the pricing model: is the consuming team charged at the platform's actual cost, at a rate that includes the platform team's operational overhead, or at a market-rate benchmark against hosted API alternatives? Each choice shapes the behaviors it incentivizes.

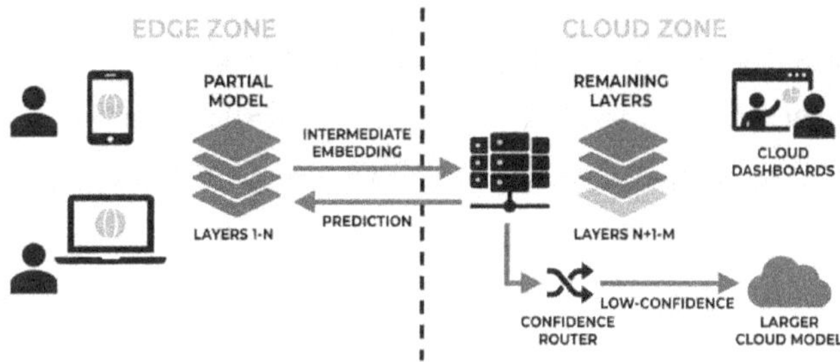

The practical implementation of both models depends on workload tagging. Every inference request must include a team identifier, a feature identifier, and a workload-type tag that the cost-attribution system can use to aggregate spend by consuming team. These tags are typically injected at the API gateway or the serving layer's request context. Tag coverage needs to be enforced at the platform boundary: requests without valid tags should either be rejected or placed in an unattributed bucket visible on the platform team's dashboard, creating an incentive for consuming teams to maintain correct tagging.

9.11 Reservation, Commitment, and Capacity Planning Strategies

Cloud providers offer substantial discounts for capacity commitments: agreements to use a specified level of resources for a one-year or three-year term in exchange for pricing thirty to sixty percent below on-demand rates. For LLM workloads with predictable baseline traffic, these commitments represent a significant cost reduction on the infrastructure component of the bill. The design challenge is matching the committed quantity to the predictable baseline without over-committing to capacity that will go unused during traffic valleys.

The capacity planning process for commitments follows a three-step pattern. First, characterize the traffic distribution: what is the minimum sustained load across the prior ninety days, and how stable is that minimum across weeks and months? The minimum sustained load is the safe commitment level, because the commitment will be fully utilized even during the quietest periods. Second, model the traffic growth rate: at what rate is the workload's request volume growing, and will the committed capacity remain right-sized for the committed term? A commitment that is correctly sized today but will be outgrown in six months results in on-demand overage costs that partially erode the savings. Third, calculate the break-even point: the committed pricing saves a fixed percentage per unit; the commitment is justified if the expected utilization over the term exceeds the utilization level at which the savings exceed the over-commitment risk.

For teams using hosted API inference, the analogous construct is a provisioned throughput reservation: a commitment to a minimum throughput level (measured in tokens per minute) at a discounted rate. The same sizing logic applies. The reservation covers the predictable baseline; on-demand capacity covers the variable fraction above the baseline. The mix should be reviewed quarterly as traffic patterns shift. A reservation that was right-sized nine months ago may be undersized if the feature has grown, or oversized if a competitor's product has displaced some usage.

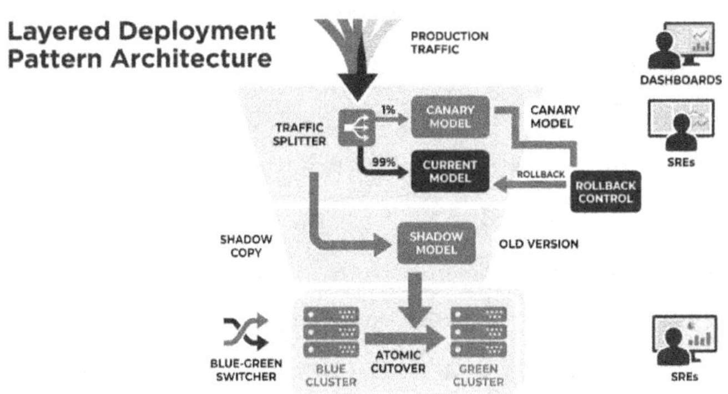

Capacity planning for self-hosted inference has an additional dimension: the hardware procurement cycle. GPU capacity is not always available on demand at the required scale; during periods of high industry demand, the lead time from reservation to deployment can extend to weeks. Teams that discover they need to scale their serving fleet in response to an unexpected traffic event may find that on-demand cloud GPU availability is constrained. Operational control involves maintaining a capacity forecast that is reviewed monthly and acted on with sufficient lead time to avoid the constrained availability window.

9.12 FOSS, Self-Hosted, and Hybrid Cost Profiles

The build-versus-buy decision in LLM serving is primarily a cost-of-scale decision. Hosted API inference offers zero operational overhead and immediate access to frontier-tier models at a per-token price. Self-hosted inference with open-weight models offers lower per-token cost at scale and full control over the serving stack. Still, it comes with GPU infrastructure management, model update cycles, and the engineering capacity required to operate the serving cluster. The economic crossover point where self-hosting becomes cheaper than the hosted API depends on the specific model tier and the team's operational efficiency.

The variables that determine the crossover are the per-token cost of the hosted API, the GPU cost per hour on the cloud provider, the throughput achievable per GPU (measured in tokens per second at the required quality level), and the team's operational cost (engineer-hours required to maintain the serving infrastructure). At low request volumes, hosted APIs are almost always cheaper because the fixed operational cost of the self-hosted fleet is not amortized across enough requests. At high request volumes, the per-token cost savings of self-hosted inference typically exceed the operational overhead. The inflection point for most teams running mid-tier open-weight models is in the range of several hundred million tokens per month. Still, this figure varies significantly with the specific hardware and the efficiency of the serving stack.

Hybrid architectures mix both options to capture the advantages of each. A common hybrid pattern routes latency-sensitive, low-volume production traffic to a hosted API (accepting the per-token premium in exchange for zero operational overhead and high reliability), while routing high-volume, latency-tolerant workloads (batch summarization, async classification, evaluation pipelines) to a self-hosted cluster. The hybrid approach also provides a fallback path. If the self-hosted cluster is unavailable due to a spot preemption event or a hardware failure, traffic can be rerouted to the hosted API as a backstop, accepting the cost increase as an availability cost rather than a steady-state expense.

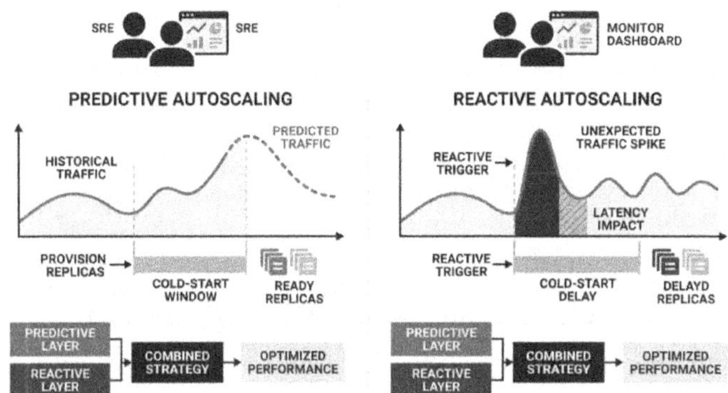

The operational cost of self-hosting is the most commonly underestimated variable in the build-versus-buy analysis. The direct costs are visible: GPU instances, networking, and storage. The indirect costs are not: the engineer-hours required for cluster provisioning, model update management, serving framework upgrades, capacity planning, on-call coverage, and compliance attestation for the serving infrastructure. A team that accurately accounts for these indirect costs often finds that the effective per-token cost of self-hosting is higher than it appears from the infrastructure bill alone, and that the crossover point is at a higher request volume than the initial estimate suggested.

9.13 A Cost Control Playbook with Tactics in Priority Order

The tactics covered in this chapter do not carry equal leverage. A quarterly cost-reduction effort that attempts all of them simultaneously will be unfocused and slow. A catalog that sequences them by expected impact, operational cost, and risk gives the team a backlog it can work through systematically. The ordering below reflects the typical leverage of each tactic across the range of production LLM workloads, with the caveat that the specific characteristics of a given workload may reorder the sequence.

Tier 1 tactics are high-impact and low in operational cost. Model-tiering and cascade routing belong in this tier: the implementation requires an offline evaluation to set the threshold and a routing layer, but the cost savings for over-modeled workloads are immediate and compounding. Exact-match caching belongs here as well: the implementation is a key-value lookup, and even modest hit rates on high-volume workloads produce meaningful savings. These tactics should be applied first, before any infrastructure change, because their leverage is highest relative to the implementation effort.

Tier 2 tactics require more engineering but deliver sustained savings. Continuous batching configuration on self-hosted serving clusters falls here: the engineering effort is a change to the serving framework, and the throughput improvement is immediate. Semantic caching is also Tier 2: the implementation requires embedding infrastructure and threshold tuning, but the hit rate expansion beyond exact-match can be substantial for workloads with paraphrastic prompt patterns. Chargeback or showback implementation belongs in Tier 2 as well, because it does not reduce costs directly but changes the behavior of consuming teams in ways that reduce costs indirectly and sustainably.

Tier 3 tactics are structural and require longer lead times. Spot instance strategy and multi-pool architecture are Tier 3 for self-hosted workloads: they require changes to the serving architecture, preemption testing, and fallback policy design. Capacity commitment negotiation is Tier 3 as well: it requires a stable traffic history and a multi-quarter commitment horizon. These tactics produce the largest absolute savings at scale, but they require the infrastructure and operational maturity of a team that has already addressed the Tier 1 and Tier 2 opportunities.

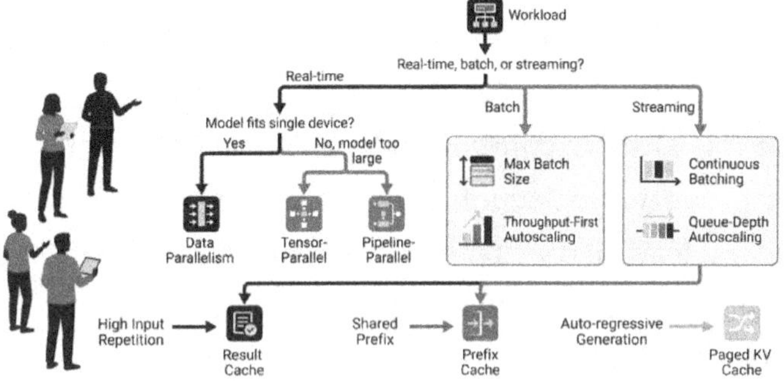

The tactic catalog becomes a living backlog when the team commits to a quarterly review cycle. At each review, the team examines its current unit economics against the targets set at the previous review, identifies the tier with the next-highest-leverage unimplemented tactic, and assigns a cost owner and a ship target. This cycle transforms cost optimization from a one-time fire drill into a predictable engineering discipline. Teams that maintain this cadence consistently find that their unit economics improve even as traffic grows, because the optimization rate keeps pace with consumption.

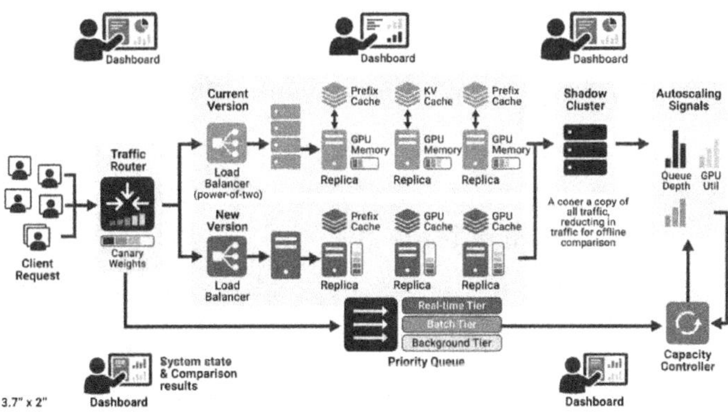

The cost control playbook closes with two principles that apply regardless of which specific tactics are in the backlog. First, cost targets must be set before traffic arrives, not after. A team that sets a cost-per-request target at launch and measures against it from day one will detect regressions immediately. A team that defers cost targets until a crisis forces the issue will always be reacting to a problem that has already compounded. Second, cost ownership must be explicit. Every production LLM workload needs a named engineer or team lead accountable for its unit economics, who reviews the cost dashboard as routinely as the latency dashboard. Without explicit ownership, cost is everyone's concern in general and no one's concern specifically, which is the organizational condition that produced the opening scenario.

9.14 Technical Checklist

Every production LLM workload has a documented cost owner and a unit-economics target reviewed at each major traffic milestone.

A cost register documents each workload's cost class (inference, fine-tuning, or both), billing mechanism, current unit cost, and amortization horizon.

Per-request cost instrumentation emits token counts, model tier, workload identifier, and team identifier for every inference call.

A cost-per-request dashboard displays a rolling time series for each workload, with baseline bands and alert thresholds configured.

Short-window alerting detects runaway agents or infinite loops before the session token budget is exhausted.

Model-tiering or cascade routing is applied to any workload where request complexity varies, and a small model can handle a meaningful fraction of requests.

Exact-match caching is in place for all workloads where the prompt repetition rate exceeds 10%.

Semantic caching is evaluated for workloads with paraphrastic prompt patterns; break-even analysis is documented before deployment.

Cache TTL policies are documented per workload and matched to the time-sensitivity of the information in the response.

Server-side continuous batching is enabled on all self-hosted serving clusters; client-side bundling is applied to all batch workloads.

Spot or preemptible capacity is used for batch inference workloads; multi-pool fallback is tested under simulated preemption before production use.

Showback or chargeback is in place at a granularity that the consuming teams can act on; tag coverage is enforced at the API gateway.

Capacity commitments are reviewed quarterly; committed levels are matched to the minimum sustained load observed over the prior ninety days.

The crossover point between hybrid and hosted costs has been calculated for each major workload, accounting for indirect operational costs.

A tactic catalog is maintained and reviewed quarterly; at least one tactic is shipped per quarter with a cost owner assigned.

9.15 Team Conversation

- What are our costs per request and per user session today, and are those numbers measured or estimated? Would the measurement hold up to a financial audit?
- Which of our production workloads are over-modeled right now, using a large or frontier model where a smaller

one would produce equivalent quality at a fraction of the cost?

- Where in our stack are we paying interactive API prices for workloads with no latency requirements? What would it take to route those workloads to a batch endpoint or a self-hosted cluster?
- Have we ever detected a runaway prompt, agent loop, or accidental model tier migration in real time, or have we only discovered these situations when the monthly bill arrived?
- Is our showback or chargeback model granular enough that individual teams can see how their decisions affect the bill and act on what they see? If not, what is the minimum change to the tagging needed to get there?
- What is the next tactic in our cost catalog, and what specific engineering or organizational blocker prevents us from shipping it this quarter?
- If our inference volume doubled tomorrow without a corresponding change in revenue, would our unit economics hold up, or would we be in a cost crisis? What is our break-even traffic level?
- How often do we review our capacity commitments against our actual traffic baseline? Who owns that review, and when did we last do it?
- What is the indirect operational cost of our self-hosted serving infrastructure when fully loaded (including on-call, upgrade cycles, compliance attestation, and capacity planning time)? Have we ever compared that number directly to the hosted API alternative?
- Is there a single named person on the team accountable for the cost-per-request of each production workload, or is

cost accountability distributed so that it effectively belongs to no one?

9.16 Key Takeaway

LLM cost is not a billing category to be reviewed monthly; it's a quarterly concern. It is an engineering variable that changes with every prompt design decision, every model selection, every serving configuration, and every workload routing policy. Teams that treat it as an engineering discipline, instrument it at the request level, assign explicit ownership, and work through a tactic catalog systematically and consistently achieve unit economics that improve as their systems mature. Teams that treat it as a finance concern discover its dynamics only after the damage is visible on the bill.

The three highest-leverage interventions are model-tiering, caching, and cost observability. Model-tiering closes the gap between the model the workload requires and the model it is using. Caching removes the inference computation entirely for repeated prompt patterns. Cost observability turns regressions from monthly surprises into real-time signals that can be investigated and resolved before they compound. None of these interventions requires new infrastructure in the conventional sense. They require measurement, discipline, and the organizational clarity of named ownership.

The playbook in this chapter is designed to be a living backlog, not a one-time project. The quarterly review cycle, the tactic catalog sequenced by leverage tier, the unit-economics targets set before launch: these are the operational habits that make cost engineering a repeatable discipline. A team that builds these habits early, before a cost crisis forces the issue, will find that cost becomes a managed variable rather than an unmanaged risk, and that the resulting operational clarity extends well beyond the billing dashboard into the team's overall understanding of how its system behaves under load.

10 Safety, Security, and Compliance

10.1 Opening Scenario

The production incident did not arrive as an alert. It arrived as an email forwarded from a customer's security team on a Wednesday afternoon, twenty-six hours after the B2B software company had shipped its AI assistant feature. The customer's penetration tester had discovered that a carefully constructed prompt could cause the assistant to repeat fragments of its system instructions, including a section that named the internal policy engine the company used for content moderation. A second variation of the same prompt produced a coherent paragraph about the assistant's underlying model family and the approximate dataset size used in a recent fine-tuning run. Neither output was catastrophically sensitive in isolation. Together, they were enough for the customer's security officer to flag the integration as a compliance risk and to request remediation within 72 hours, or face contract suspension.

The engineering team's first instinct was to patch the system prompt by adding a stronger instruction not to reveal internal information. The security team's first question was more uncomfortable. If a prompt injection could extract system instructions, what else could it extract? Could it be used to leak other users' conversation context? Could it be prompted into producing outputs that violated the company's content policy in ways that would surface in audit logs? Could it be combined with a crafted payload that caused the model to invoke an API tool it was not supposed to invoke? The scope of the question expanded within hours from a single prompt to a surface-area audit of every interface the model touched.

The remediation required more than a patched prompt. The team implemented input sanitization that detected and quarantined injection-structured payloads before they reached the model. They added an output classifier that scored every response against a policy schema before it left the system. They instrumented the assistant's tool invocations so that every API call was logged with the prompt hash that triggered it, enabling forensic reconstruction of any future incident. They wrote a threat model document for the first time, naming each attack surface, the control that addressed it, and the team member who owned the control. Seventy-one hours after the customer's email arrived, the security officer received a written remediation report, a copy of the threat model, and a scheduled red-team engagement. The contract survived. The architecture that resolved the crisis became the foundation the team used for every subsequent LLM feature.

This chapter is that architecture, generalized. It covers the threat surfaces LLMs introduce that classical application security frameworks do not address, the layered defense model that contains blast radius when any single control fails, and the compliance posture that lets a team demonstrate due diligence to the security, legal, and product leadership who will evaluate it. The patterns here are operational, not academic. Each one answers the engineer's question: what do I build, and how do I know it is working?

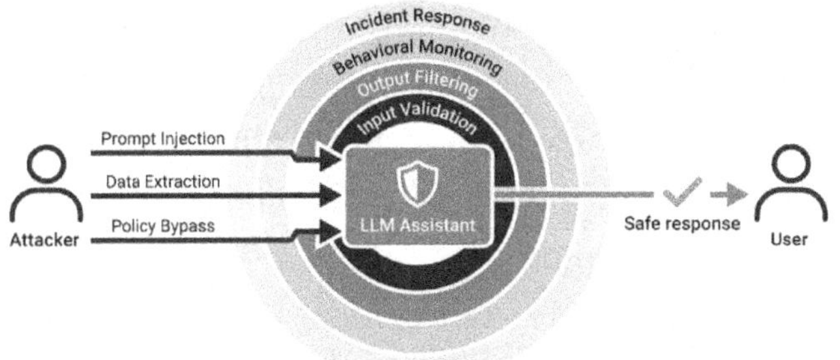

Layered defense contains blast radius when any single control fails.

10.2 Why It Matters

LLM features expose attack surfaces that do not exist in classical software. A traditional web endpoint receives structured input, executes deterministic logic, and returns a structured response. An LLM endpoint receives natural language, executes probabilistic inference over billions of parameters, and returns natural language. That shift from structured to unstructured, and from deterministic to probabilistic, opens a class of vulnerabilities that no amount of SQL injection filtering or XSS escaping will address. The AppSec playbook that the team has refined over the years does not transfer. Prompt injection, hallucination, model memorization, context leakage, and policy bypass through adversarial phrasing are new problem shapes that require new defenses.

The organizational stakes are equally new. When a classical application has a security incident, the blast radius is usually well-bounded: a specific endpoint was compromised, a specific dataset was accessed, a specific action was taken. When an LLM feature has a security incident, the blast radius is harder to characterize because the model's behavior is emergent. The same model that extracts system instructions under one prompt variation may produce harmful content under a different one, or invoke a tool chain in an unintended sequence under a third. Security teams that review LLM features for the first time often find that the team lacks a threat model, an audit trail of model outputs, and behavioral monitoring to detect anomalous output patterns at runtime. That gap is a regulatory and reputational liability, and closing it is now part of shipping responsibly.

The compliance dimension compounds the engineering challenge. Sector regulations that govern data handling, such as those covering health records, financial transactions, or personal data in specific jurisdictions, do not carve out exceptions for AI-generated responses. If a model produces a response that contains a patient's medical history, or a customer's account balance, or a regulated individual's personal information, the organization is liable for that disclosure under the same frameworks that govern direct database queries. The compliance posture required for a production LLM is not additive to the classical posture: it is a superset that requires new controls on top of every existing one. The teams that treat this chapter's checklist as a release gate, not an afterthought, are the ones that ship features their security and compliance partners can stand behind.

10.3 Adversarial Inputs and Prompt Injection Defense

Prompt injection is the defining attack class for LLM systems. In its direct form, a user submits a payload that attempts to override the model's system instructions: a crafted sentence that begins with a phrase like 'ignore previous instructions' and substitutes the attacker's goal for the application's intended behavior. In its indirect form, the malicious instruction is embedded in content the model retrieves and processes rather than content the user submits directly: a document the model is asked to summarize, a webpage the model is instructed to read, or a database record the model is asked to transform. Indirect injection is more dangerous in production because it does not require the attacker to have direct access to the prompt interface.

The defense is layered because no single control is sufficient on its own. Input sanitization at the perimeter catches the most obvious patterns: payloads that contain known injection phrases, unusual Unicode representations, or structural anomalies that suggest prompt manipulation. Sanitization is necessary but not sufficient, because the injection surface is infinite and any blocklist is incomplete. The second layer is instruction isolation, implemented through architectural separation between the system prompt and user-supplied content. When the model is constructed to treat these as structurally distinct rather than concatenated strings, the surface area for override attacks is reduced. Context-fencing patterns extend this principle by limiting what the model is permitted to reference: a model that is only allowed to reason over a specific retrieved context cannot be prompted to reason over its training data or its system instructions as if they were user-supplied content.

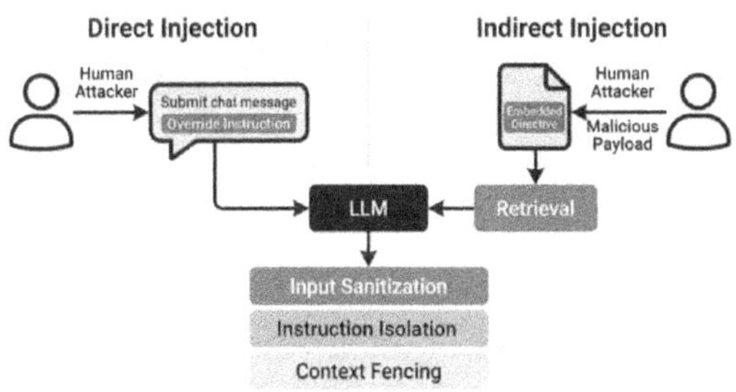

Indirect injection through retrieved content is harder to detect than direct user input attacks.

Behavioral monitoring is the third layer. When sanitization and isolation are in place, behavioral monitoring catches the injections that bypass both. A monitor that tracks the statistical distribution of output types across requests will detect anomalies: a sudden spike in responses that contain system-prompt fragments, a cluster of requests that trigger tool invocations outside normal parameters, or a sequence of prompts that systematically probe the model's response boundaries. These signals are not false-positive-free, but they provide the detection surface that the other two layers cannot offer.

The blast-radius containment principle governs the whole stack. Each layer is designed to fail independently, so that bypassing one layer does not bypass all three. A payload that evades input sanitization encounters instruction isolation. A payload sophisticated enough to exploit a structural flaw in instruction isolation still faces behavioral monitoring. The goal is not an impenetrable perimeter, which does not exist for LLM systems, but a layered architecture where any single failure has bounded consequences and leaves detectable evidence.

10.4 Hallucination Detection and Mitigation

Hallucination is simultaneously a quality, safety, and compliance problem. When a model produces a plausible but factually incorrect output, the user experiences a quality failure. When the incorrect output involves a safety-relevant claim, such as a medical dosage, a legal interpretation, or a financial figure, the quality failure becomes a safety failure. When the incorrect output is presented in a regulated context, and a record of the interaction is available, the safety failure becomes a compliance finding. The appropriate response to hallucination is not to treat it as an acceptable model characteristic and move on. It is to build detection and mitigation into the serving path.

Grounding via retrieval is the most reliable mitigation for factual hallucination. When the model's output is constrained to facts that appear in a retrieved context, the space of possible hallucinations is reduced to the space of misquotation and misattribution, which is smaller and easier to detect than unconstrained fabrication. Faithfulness scoring operationalizes this constraint: after the model produces a response, a secondary scoring pass checks whether each claim in the response can be traced to a span in the retrieved context. Claims that cannot be traced are flagged for review, filtered, or cause the system to return a structured refusal rather than an ungrounded assertion.

Hallucination Detection and Mitigation Pipeline

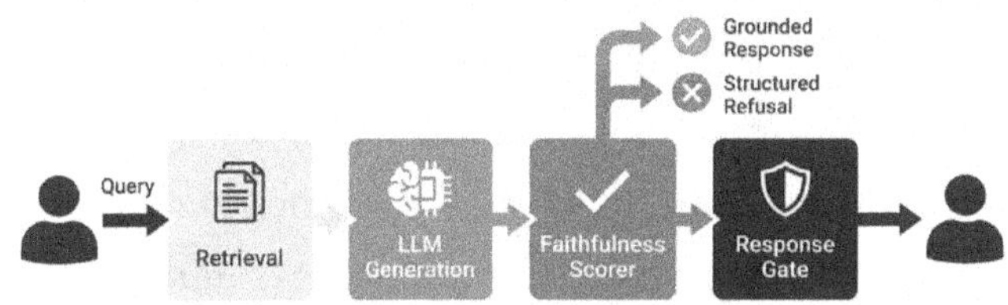

Faithfulness scoring after generation is the operational control that makes retrieval grounding enforceable.

Refusal training addresses a different hallucination mode: the model that confidently answers questions it should decline to answer. A model that has been fine-tuned or prompted to recognize the boundaries of its competence and to express uncertainty or decline rather than fabricate reduces the rate of confident hallucinations in edge cases. Refusal training is most effective when the training or prompt examples are drawn from the specific domain the model operates in, because domain-generic uncertainty patterns do not transfer well to domain-specific edge cases.

Structured-output patterns make hallucinations easier to catch downstream. When the model is constrained to produce JSON, YAML, or another structured format rather than free text, validators can check the output against a schema before it is consumed. A response that contains a field with a value that violates the schema's type constraint or enumeration is automatically flagged. Free-text hallucinations are harder to validate because their failure mode is semantic rather than structural. Wherever the application permits structured output, preferring it over free text is a low-cost mitigation that pays dividends across the entire quality and safety stack.

10.5 Output Filtering, Guardrails, and Policy Enforcement

Input defenses reduce the probability of unsafe outputs but cannot eliminate them. The model may produce an unsafe output in response to a benign input, because the relationship between input and output in a large language model is not a deterministic function but a probability distribution. Output filtering is the defensive layer that catches what input sanitization misses: a policy-aware classifier that evaluates every response against the organization's content and safety policies before returning it to the caller.

Policy-aware classifiers are the core mechanism. A classifier trained or fine-tuned on the organization's specific policy schema, rather than a generic content safety taxonomy, will perform better on the edge cases that matter most. A consumer-facing assistant has a different policy than an internal developer tool, and a healthcare assistant has a different policy than an e-commerce recommendation engine. Deploying a generic classifier and hoping it matches the organization's policy is a source of both false positives, which degrade user experience, and false negatives, which constitute safety failures. The investment in policy-specific classifier training or fine-tuning is the difference between a guardrail that holds and one that is bypassed by the first adversarial prompt the team encounters.

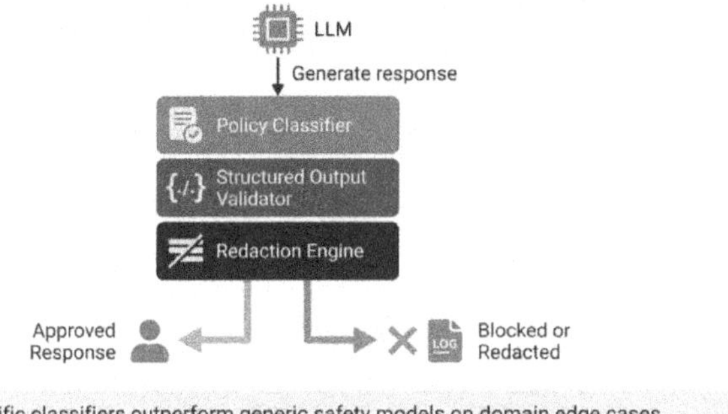

Policy-specific classifiers outperform generic safety models on domain edge cases.

Redaction patterns address the subset of unsafe outputs that contain sensitive data rather than policy-violating content. A response that inadvertently includes a user's personal identifier, an internal system name, or a credential fragment that was present in a retrieved document must be redacted before it reaches the caller. Redaction engines operate on a set of patterns and entity recognizers: regular expressions for structured formats such as social security numbers or API keys, named-entity recognition for person and organizational names, and custom patterns for organization-specific sensitive tokens. Redaction is a last line of defense, not a substitute for the data minimization patterns that prevent sensitive content from entering the retrieval context in the first place.

Runtime enforcement patterns close the loop between detection and action. When a classifier scores a response above a policy threshold, the enforcement layer must take a defined action: block the response and return a structured refusal, redact the flagged spans and return the remainder, escalate to a human reviewer, or log the event and return the response with a risk annotation that a downstream consumer can act on. The choice among these actions is a policy decision, not a technical default. Teams that have not explicitly defined their enforcement actions will find their classifiers operating in detection-only mode indefinitely, producing signals that no one acts on.

10.6 Access Control and Authentication for LLM APIs

LLM API endpoints require the same access-control rigor as any other sensitive system interface, and several additional controls that the unique characteristics of LLM interfaces necessitate. The additional controls address three failure modes not present in classical API security: credential extraction via prompts, per-tenant context leakage, and rate-limit abuse enabled by the high cost of LLM inference.

Credential extraction through prompts occurs when a model has been provided with credentials, API keys, or connection strings in its system prompt or retrieved context, and an adversarial prompt causes it to repeat those credentials in its response. The control is structural: credentials should never appear in any content processed by the model. Configuration injection, in which the model receives dynamic configuration via a structured parameter rather than as text in the prompt, prevents credential leakage through this channel. When credentials must appear in a retrieval context for operational reasons, that context should be filtered before it reaches the model, with credentials replaced by opaque references that the serving infrastructure resolves after the model's response is generated.

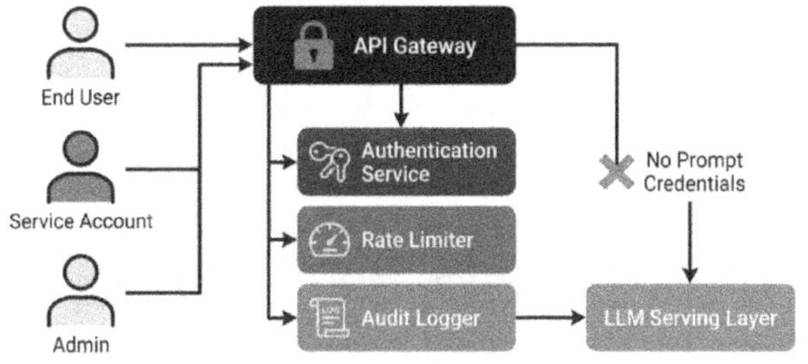

Workload identity, rather than long-lived API keys, is the preferred authentication pattern for service-to-service LLM API calls. A workload identity framework issues short-lived credentials that are scoped to specific operations and automatically rotated, removing the operational burden of key rotation and eliminating the credential-lifespan risk that long-lived keys create. For human callers, user-identity-based access control with appropriate scoping, such as restricting a user's access to models and data sources relevant to their role, limits both the blast radius of compromised credentials and the exposure surface for prompt injection attacks that attempt to access cross-tenant data.

Per-tenant rate limits serve both security and cost-control functions. A user or service account that is not rate-limited can issue an arbitrarily large number of inference requests, creating two risks: financial exposure from unexpected cost spikes, and denial-of-service risk to other tenants sharing the same serving infrastructure. Rate limits scoped to the tenant, rather than to the endpoint globally, ensure that one tenant's traffic cannot starve another's. When combined with anomaly detection on per-tenant request patterns, rate limits also provide a signal for detecting automated adversarial probing, which typically manifests as a burst of structurally similar requests at a rate higher than any human user would produce.

10.7 Data Privacy, PII Handling, and Memorization Risk

LLMs pose privacy risks that differ from those of classical data systems in two important ways. First, the model's inference process is not auditable in the way a database query is: no query log shows exactly what data was accessed to produce a given output. Second, the model may have memorized specific data points from its training corpus, and a carefully constructed prompt can cause it to reproduce that memorized data even when the data is not present in the retrieval context for the current request. Both risks require controls that have no direct analog in classical data privacy frameworks.

PII detection and redaction of inputs are the first controls. Before a user's input reaches the model, a PII detector scans it for personal identifiers: names, addresses, phone numbers, email addresses, government identifiers, health-related terms, and financial account numbers. Detected entities are either redacted before the input is passed to the model, replaced with type-preserving synthetic equivalents that allow the model to process the request without seeing the actual identifier, or flagged for a policy decision about whether to proceed. The choice among these strategies depends on the use case: a document summarization feature can usually redact before passing; a support assistant that needs to resolve account information requires a more nuanced approach that keeps identifiers accessible to the serving infrastructure without embedding them in the raw prompt.

PII Detection and Redaction in the LLM Request Pipeline

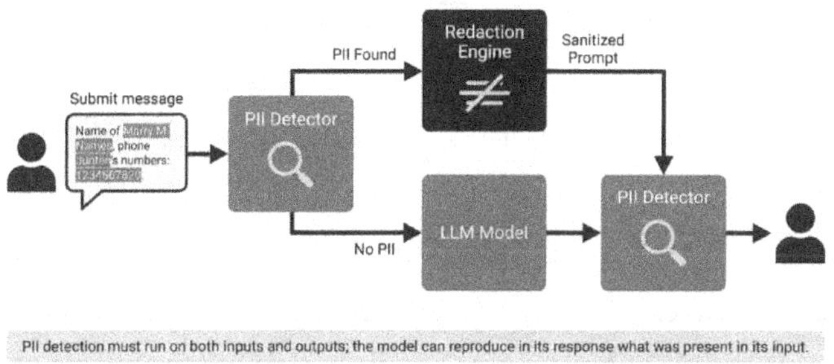

PII detection must run on both inputs and outputs; the model can reproduce in its response what was present in its input.

PII detection and redaction in outputs are equally necessary. The model can reproduce content present in its input, in its retrieval context, or in its training data. Output-side PII detection applies the same entity-recognition pipeline to responses before they are returned to the caller. This is computationally redundant with input-side detection but not logically redundant: the model can synthesize PII from multiple partial signals that would not individually trigger input-side detection.

The risk of memorization in fine-tuned models warrants specific attention. When a model is fine-tuned on a dataset containing real personal data, it can memorize and later reproduce specific data points from that dataset. The risk is proportional to the frequency with which a specific data point appears in the training data: rare occurrences are less likely to be memorized than repeated ones. Data minimization in fine-tuning datasets, specifically removing or anonymizing personal data before using a dataset for fine-tuning, is the primary control. Differential privacy techniques, which add calibrated noise to gradient updates during fine-tuning, provide a formal privacy guarantee at the cost of some model performance. The appropriate trade-off between privacy protection and model performance is a policy decision that should be made explicitly, with input from both the privacy and the engineering functions, rather than by defaulting to no protection.

10.8 Logging Policies and Audit Trails for LLM Interactions

Audit logging for LLM interactions is more complex than audit logging for classical APIs because the content of LLM interactions, the prompts and responses, is itself potentially sensitive. A log containing the full text of every user prompt may include PII, privileged business information, or confidential queries that should not be retained indefinitely or accessible to the same operators who manage the serving infrastructure. The design of the logging system must balance two competing requirements: the completeness needed to detect security incidents and satisfy regulatory audit requirements, and the data minimization needed to avoid creating a second data-exposure surface in the logs themselves.

Redacted logs are the standard pattern. Rather than logging full prompt and response text, the logging system stores a structured event record that contains: a hashed prompt identifier that can be used to correlate events across systems without storing the prompt content, the model identifier and version, the policy classifier scores for the request and response, any tool invocations triggered and their parameters, the response latency and token counts, and any anomaly flags raised by behavioral monitoring. This record supports forensic reconstruction of an incident without requiring the retention of full prompt text. When full prompt content must be retained, such as for regulatory compliance in a sector that requires verbatim records of AI-generated advice, it should be stored in a separately access-controlled store with its own retention and deletion schedule.

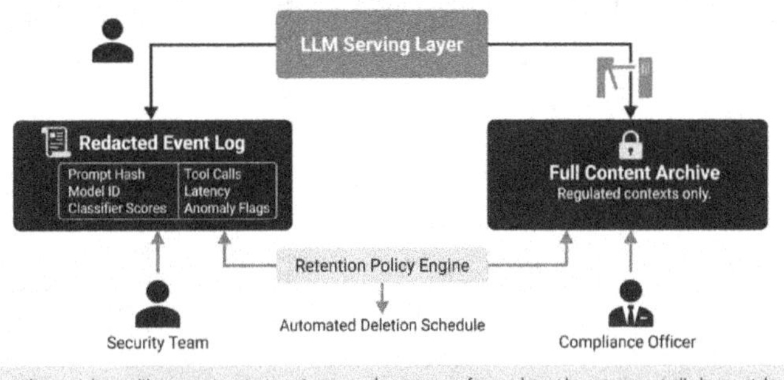

Retention policies for LLM interaction logs must account for two distinct requirements that often have different time horizons. Security incident detection requires retaining enough log history to reconstruct the sequence of events that preceded a detected anomaly, typically measured in weeks to months. Regulatory compliance may require retaining records of AI-generated advice or decisions for years, depending on the sector and jurisdiction. These requirements should be encoded in the retention policy as distinct categories with distinct schedules, rather than applying a single retention period to all log data. When a regulatory audit requires access to logs, a process for producing the required records without exposing the full log store to the auditor is both a practical necessity and a privacy protection for users whose interactions are incidentally captured.

Access controls on logs must be as rigorous as access controls on the data the logs describe. Logs that capture user queries and model responses may be the most sensitive artifact in the system, because they provide a complete record of what users asked and what the system said. A policy that allows broad read access to logs for operational convenience undermines the privacy protections applied to the raw data. Role-based access controls, with specific log-access roles separate from the roles that grant access to the serving infrastructure, are the minimum viable posture. Audit logs on the log access system itself close the loop: if the logs are ever accessed inappropriately, there is a record.

10.9 Adversarial Red-Teaming and Continuous Testing

Red-teaming for LLM systems has become a standard practice because the attack surface of a deployed model cannot be fully characterized by static analysis or unit testing. The model's behavior is a function of its training, its prompt context, and the specific input it receives. The space of inputs is unbounded, and the set of inputs that produce unsafe outputs is not predictable from the model's architecture alone. Red-teaming, the systematic attempt to find inputs that produce policy-violating, harmful, or otherwise unacceptable outputs, is the discovery process that reveals the actual failure modes of the deployed system.

Automated adversarial probing is the foundation of a scalable red-team program. A suite of automated probes covers the known attack taxonomies: direct prompt injection, indirect injection via retrieved content, jailbreak patterns, sensitive-topic elicitation, system instruction extraction, PII elicitation, and tool-misuse prompts. These probes are run against every significant model update, system prompt change, and retrieval context change as part of the CI pipeline. Automated probing does not replace human red-teaming but establishes a regression baseline: any change that increases the model's vulnerability on the automated suite is blocked before deployment, and human red-teamers focus their time on novel attack patterns and domain-specific edge cases that the automated suite has not yet incorporated.

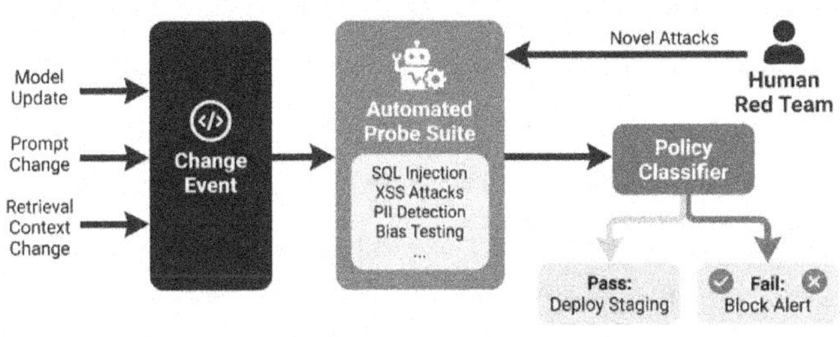

Automated probes set the regression baseline; human red-teamers extend the suite with novel attack patterns.

The red-team cadence should be documented and enforced. Ad hoc red-teaming that occurs informally is not sufficient evidence of due diligence for a compliance audit. A documented cadence specifies how frequently automated probing runs, how frequently human red-team engagements occur, what the process is for incorporating new findings into the automated suite, and who is responsible for reviewing findings and deciding on remediation. The cadence documentation is itself a compliance artifact: it demonstrates that the organization has a systematic safety-testing process, not just a one-time evaluation at launch.

Model regression testing extends the red-team program into the standard CI/CD pipeline. Every model update, including fine-tuning runs and changes to the retrieval index, should be tested against a safety regression suite before being promoted to production. The safety regression suite differs from the functional test suite in that it is designed to detect degradation in safety behaviors rather than in capability metrics. A model update that improves performance on a capability benchmark but increases the rate of policy violations on the safety regression suite should not be promoted regardless of the capability improvement. Separating the safety gate from the capability gate ensures that capability pressure does not silently erode safety posture.

10.10 Watermarking, Provenance, and Content Authenticity

As LLM-generated content becomes indistinguishable from human-generated content in most surface-level evaluations, the ability to assert provenance, to answer the question of whether a specific model generated a specific piece of text, has become operationally and legally significant. Organizations that use LLMs to produce content that is subsequently attributed to human authors, presented as factual reporting, or submitted in regulated contexts where the source of the content is material to its validity, need the ability to assert or deny AI generation with a degree of confidence that informal inspection cannot provide.

Soft watermarking embeds a statistical signal in the model's output distribution during generation. By biasing the model's token sampling toward tokens that encode a detectable pattern without materially affecting the output's quality or coherence, the watermarking system creates a signal that a detection algorithm can identify in the output text, even after normal editing. Soft watermarks are not robust to aggressive paraphrasing or translation, but they are sufficient for many provenance-verification use cases in which the text is reproduced substantially intact. The detection algorithm requires access to the same key used during generation, limiting detection capability to parties who hold the key.

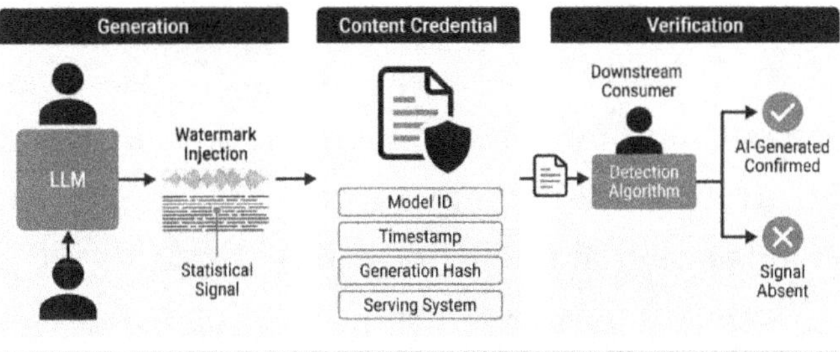

Content credentials provide a complementary provenance mechanism that does not depend on statistical signal embedding. A content credential is a structured metadata record attached to the generated content at the point of production, containing the model identifier, the generation timestamp, a cryptographic hash of the generation parameters, and a signature that can be verified against the organization's public key. Unlike a soft watermark, a content credential is not embedded in the content itself and can be stripped by any intermediary who chooses to remove it. Its value lies in audit trails and controlled-distribution contexts, where intermediaries are trusted, and the credential is expected to travel with the content.

Propagation through downstream systems is the hardest operational challenge for any provenance framework. Content that is generated with a watermark or credential and then copied into a document management system, a communication platform, or a web publishing tool will lose its provenance signal if those downstream systems do not understand and preserve the credential format. Organizations that intend to rely on provenance controls for compliance or legal purposes should audit their content workflow to identify where provenance signals are preserved and where they are dropped, and should treat the preservation of provenance signals as a requirement when evaluating new downstream systems.

10.11 Regulatory Compliance Frameworks

The regulatory landscape for AI systems is no longer a future concern. The EU AI Act's risk-tier framework, sector-specific data-handling regulations, and the emerging ISO standards for AI management systems are active compliance obligations for organizations deploying LLM features in regulated contexts. Treating compliance as a design deliverable, rather than a post-hoc audit, is the difference between a system that passes a compliance review and one that requires a remediation sprint before it can.

The EU AI Act's risk classification determines the compliance obligations a system must satisfy. Systems classified as high-risk, a category that includes AI used in employment decisions, credit decisions, law enforcement, education, and critical infrastructure, face mandatory conformity assessments, technical documentation requirements, human oversight requirements, and ongoing monitoring obligations before they can be deployed in the EU market. LLM systems that do not fall into a high-risk category still face transparency obligations if they interact with natural persons who may believe they are speaking with a human. Mapping a system to the appropriate risk tier is the first compliance design decision and determines the entire set of obligations that follow.

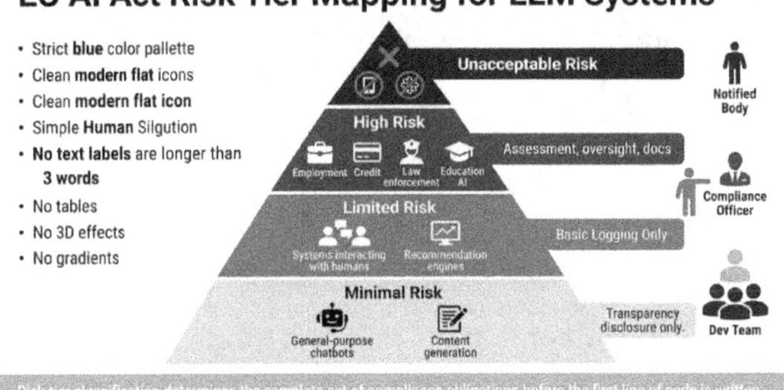

Sector-specific regulations impose additional obligations that build on framework-level requirements. In healthcare contexts governed by health data protection rules, LLM systems that process patient information are subject to the same access controls, audit requirements, and breach notification obligations as other systems that process protected health information. In financial services contexts, LLM systems used for customer-facing advice or credit decisions may be subject to explainability requirements, fair lending obligations, and model risk management guidance, including documentation of model limitations and testing for disparate impact. In any sector where records of AI-generated decisions must be retained for regulatory review, the logging and retention architecture described in an earlier section of this chapter is not optional.

ISO standards for AI management, particularly those addressing AI risk management, AI transparency, and AI governance, provide a framework for demonstrating due diligence that regulators recognize across multiple jurisdictions. Alignment with these standards is voluntary in most contexts but is increasingly requested by enterprise customers during vendor security assessments and by sector regulators as evidence of a systematic approach to AI safety. Treating standard alignment as a design input and documenting the mapping between the system's controls and the standard's requirements creates the compliance artifact that both internal and external reviewers need.

10.12 Incident Response for Model Failures

LLM incidents differ from classical software incidents in ways that require a specifically adapted incident response playbook. In a classical software incident, the failure mode is usually deterministic: a specific bug causes a specific error under specific conditions. The fix is typically a code change, a configuration update, or a rollback to a previous version, and the incident is closed when the erroneous behavior stops. In an LLM incident, the failure mode is often probabilistic and context-dependent: a specific class of prompts produces an unacceptable class of outputs at a rate that crosses a threshold, but every input in the class does not trigger the behavior. The fix may require a model change, a prompt change, a classifier update, a retrieval index change, or some combination of all of these, and the incident is not closed until the team can demonstrate through systematic testing that the failure rate has been reduced to an acceptable level.

Detection is the first stage of the LLM incident response playbook and is where the most important difference from classical incident response lies. Classical incidents are often detected through hard errors: exceptions, 5xx responses, and latency spikes. LLM incidents are often detected through soft signals: a spike in policy classifier scores, an increase in user-reported complaints, an anomaly in the distribution of output types, or a red-team finding that a new attack pattern is bypassing existing defenses. The monitoring infrastructure described earlier in this chapter is the detection system for LLM incidents. A team that has not built that infrastructure will detect LLM incidents through customer complaints rather than operational signals, meaning the incident has already been externalized before the team knows it exists.

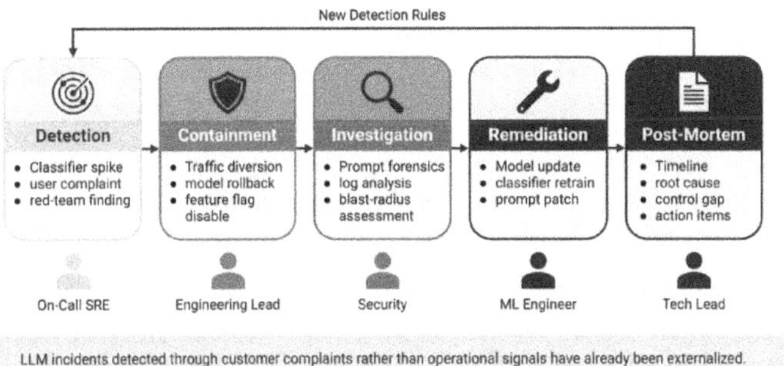

Containment in an LLM incident requires tools that are designed before the incident occurs. The ability to divert traffic from a model to a fallback, turn off specific features through a feature flag, roll back to a previous model version, or increase classifier sensitivity to reduce the false-negative rate must all be exercised under the time pressure of an active incident. Teams that are building these controls for the first time during an incident will be slower and more error-prone than teams that have rehearsed the containment procedures in advance. The incident response playbook should name each containment tool, the procedure for invoking it, and the team member with the authority and access to invoke it.

Post-mortem practice for LLM incidents should explicitly address the control gaps that enabled the incident. A post-mortem that concludes with 'the model produced an unexpected output' without identifying the specific control that should have prevented or detected it, and without assigning a remediation action to close the gap, does not improve the organization's safety posture. The five sections of a mission-aligned LLM post-mortem are: the incident timeline, the root cause in terms of which control failed or was absent, the blast radius assessment, the remediation actions with owners and deadlines, and the new detection or prevention rules added to the monitoring and red-team suite as a result of the incident.

10.13 A Safety Checklist for Production LLM Systems

The patterns in this chapter reduce to a concrete checklist that both security and product leadership can sign off on before a feature ships. A checklist signed by both functions, rather than by the engineering team alone, is the operational definition of safety that has been reviewed by the people who will be accountable for any safety failure. The checklist is a release gate, not a post-launch aspiration.

The checklist should be treated as a living document. New attack patterns, new regulatory requirements, and lessons from post-mortems will all require additions. The version of the checklist that exists at launch is not the version that should exist six months later. Assigning ownership of the checklist to a specific team member, with a documented review cadence, ensures it evolves with the threat landscape rather than becoming a static artifact that the team signs without reading.

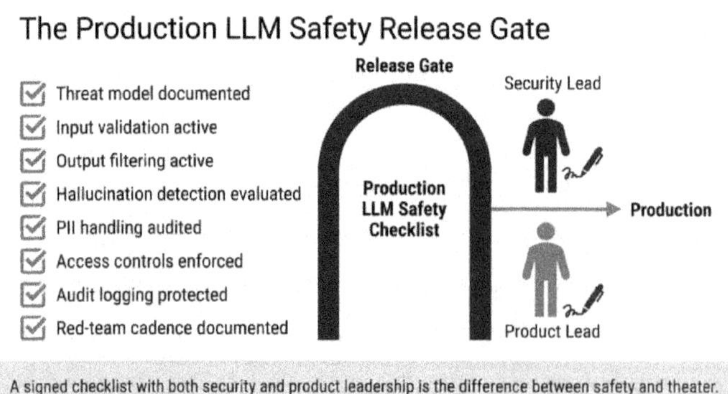

A signed checklist with both security and product leadership is the difference between safety and theater.

The items on the checklist correspond directly to the controls described in this chapter. A threat model exists and has named owners. Input validation, output filtering, and behavioral monitoring are all in place and have been tested. Hallucination detection has been evaluated for the specific use case and is active when the feature is required. PII handling has been audited end-to-end, including in fine-tuning data and in retrieval contexts. Access controls and per-tenant rate limits are enforced. Audit-trail logging is in place, protected by its own access controls, and has a documented retention and deletion schedule. Adversarial red-teaming is run on a documented cadence, and findings are incorporated into the automated probe suite. An LLM-specific incident response playbook exists, names specific containment tools and their owners, and has been exercised or rehearsed in the last quarter.

10.14 Technical Checklist

1. A documented threat model exists for the LLM feature, with named owners for each identified attack surface and control.
2. Input validation is in place, covering injection-structured payloads, Unicode anomalies, and known attack patterns, and is enforced before any user content reaches the model.
3. Instruction isolation separates system prompts from user-supplied content architecturally, not just by convention.
4. Output filtering with a policy-aware classifier is active on every response path before content is returned to the caller.
5. Hallucination detection, whether through faithfulness scoring, structured-output validation, or refusal training, has been evaluated and deployed where the feature requires factual grounding.
6. Redaction patterns for PII and sensitive tokens are active on both inputs and outputs.

7. PII handling has been audited end-to-end, including in fine-tuning datasets, retrieval contexts, and prompt templates.
8. Access controls use scoped credentials or workload identity rather than long-lived shared API keys.
9. Per-tenant rate limits are enforced, and anomaly detection is active on per-tenant request patterns.
10. Credentials, connection strings, and sensitive configuration values do not appear in any content processed by the model.
11. Audit-trail logging captures structured event records for every LLM interaction, with access controls on the log store that are separate from those on the serving infrastructure.
12. Log retention and deletion schedules are documented and enforced, with separate schedules for security logs and regulated-content archives.
13. Adversarial red-teaming runs on a documented cadence, covering prompt injection, jailbreak patterns, sensitive-topic elicitation, and tool misuse.
14. A safety regression suite is part of the CI pipeline and blocks promotion of any model or prompt change that degrades safety behavior.
15. An LLM-specific incident response playbook exists, names containment tools and their owners, and has been rehearsed or exercised within the last quarter.
16. Regulatory compliance mapping is documented, with the system's risk tier identified and the obligations for that tier mapped to specific controls.
17. Content provenance requirements, if applicable, have been evaluated, and watermarking or content credential patterns have been implemented where required.

18. The checklist was reviewed and signed by both security and product leadership before the feature was promoted to production.

10.15 Team Conversation

1. Which of the LLM-specific attacks described in this chapter could a junior attacker run against our system today using only publicly available information about our feature, and what is our current detection coverage for each?
2. Where does our current hallucination story rely on the model behaving correctly rather than on a structural guardrail that enforces correctness, and what is the blast radius if the model's behavior changes in the next update?
3. Have we ever audited our prompts, logs, retrieval contexts, or fine-tuning datasets for PII? If we found PII in any of them, what would our notification and remediation process look like?
4. When did we last run an adversarial red-team engagement against our LLM features? What was the most surprising finding, and has that finding been incorporated into our automated probe suite?
5. Do our audit trails meet the standards our regulators and enterprise customers would expect in a compliance review, or do they meet the standards our team is comfortable with, and are those the same standards?
6. If we received a customer security escalation about our LLM feature today, how long would it take us to produce a remediation report and a threat model document, and who would write them?
7. Which compliance framework or regulation is most likely to catch us off guard if a regulator or enterprise customer requested a compliance review of our LLM feature next month?

8. Has our incident response playbook for LLM failures been rehearsed in a tabletop exercise? If not, what is preventing us from scheduling one?
9. Where in our deployment pipeline does safety regression testing happen, and is it possible for a model update or prompt change to reach production without passing a safety gate?
10. If we had to explain our complete LLM safety architecture to a new security team member in thirty minutes, what are the three gaps in our current posture that we would have to disclose?

10.16 Key Takeaway

LLMs introduce attack surfaces that classical application security frameworks were not designed to address. Prompt injection, model memorization, hallucination, and context leakage are not edge cases or future concerns: they are present in every deployed LLM feature and require controls that the team must design and build, not inherit from existing security infrastructure. Teams that ship LLM features with a complete layered defense, covering input validation, output filtering, behavioral monitoring, access control, and incident response, are not doing more than teams that skip these controls. They are doing the minimum required to operate responsibly at a threat surface that their customers, regulators, and security partners are actively evaluating.

Safety is not a property of the model; it is a property of the system. A safe model behind an unsafe serving architecture, an unmonitored retrieval pipeline, or an unprotected audit trail is not a safe system. The layered defense model in this chapter is designed around the recognition that any single control will eventually be bypassed, and that the goal is to ensure each layer fails independently and leaves detectable evidence when it does.

Continuous behavioral monitoring, red-team testing on a documented cadence, and a post-mortem practice that closes the loop between incidents and improved detection are the operational disciplines that sustain a safety posture over the lifetime of a feature, not just at launch.

The compliance posture a team carries into production is visible to every enterprise customer, regulator, and security reviewer who evaluates the feature. A signed-off checklist with both security and product leadership, a documented threat model with named owners, and an exercised incident response playbook are the artifacts that distinguish a team that has built safety responsibly by design from a team that has built safety as theater. The chapter's technical checklist is the release gate. The team conversation questions are the pressure test. The key discipline is to treat both as mandatory rather than aspirational every time a new LLM feature ships.

11 Observability and Monitoring for LLMs

11.1 Opening Scenario

The question came in on a Wednesday morning via a Slack message from the product director: 'Is the assistant getting better or worse?' The team responsible for a customer-facing support assistant that had been live for six months did not have an immediate answer. Latency was stable. Error rates were flat. Request volume was up. By every signal on the operational dashboard, the system looked healthy. The team sent back a reassuring reply and opened a ticket to investigate.

Two weeks of investigation told a different story. A prompt change, shipped three months earlier, had quietly shifted the assistant's tone from concise and factual to verbose and hedged. Users were regenerating responses at nearly twice the rate they had been before the change, but the regenerate signal was not wired into any dashboard the team monitored. A thumbs-down widget existed in the interface, but the feedback it collected was being written to a database table that no alert or scheduled report referenced. An automated evaluator had been set up at launch to score faithfulness. Still, it had drifted as the upstream model was updated, and no calibration process had ever been scheduled to catch that drift.

The team counted the cost. Three months of degraded quality had reached tens of thousands of users. A customer satisfaction survey from the previous month showed a statistically significant drop in ratings for the assistant category, a fact the customer experience team had flagged in a different Slack channel, but the engineering team did not follow up. The fix required a full observability rebuild: quality service-level indicators added alongside operational ones, semantic drift detection wired into production traffic, user feedback signals routed to the same alert channel as latency pages, and a post-mortem process that captured model-specific root-cause categories. The rebuild took four weeks. The regression cost the team three months of trust.

This scenario is not unusual. It is the default outcome when a team instruments an LLM system the same way it would instrument a traditional API. Latency and error rate are necessary but nowhere near sufficient. The failure mode for an LLM is not a five-hundred status code. It is a response that is subtly wrong, quietly off-policy, or slowly diverging from what the system was designed to produce. Catching those failures requires a purpose-built observability layer that treats quality as a first-class signal. This chapter is the playbook for building that layer.

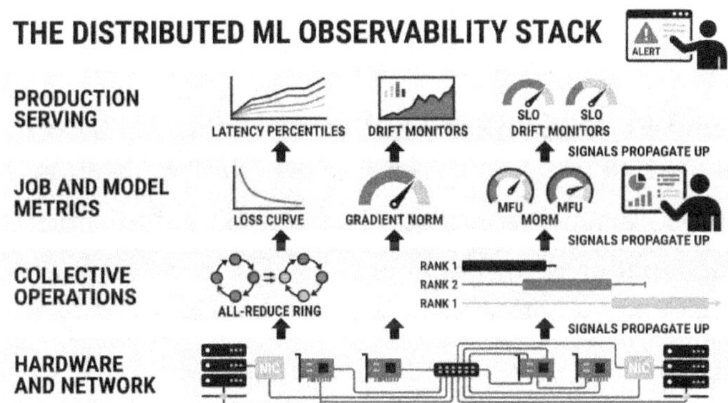

11.2 Why It Matters

Observability is not optional at scale. It is the mechanism by which an engineering team maintains operational clarity over a system it cannot fully control. For traditional software, that clarity comes from deterministic outputs: a function either returns the correct value or it raises an exception. For an LLM, correct is a spectrum. The model can return a syntactically valid, structurally complete, grammatically perfect response that is factually wrong, off-policy, or simply less useful than it was last week. No status code captures that. No latency histogram surfaces it. Only purpose-built quality signals, instrumented alongside the operational stack, will catch it in time.

The workflow-level impact of poor observability compounds over time. A quality regression that goes undetected for three months is not three times worse than one that goes undetected for one month. It is often ten times worse because the feedback loops that would have corrected the regression early were never closed. Users adapt to a degraded experience, workarounds proliferate, and by the time the regression is detected, the team faces both a technical fix and a trust deficit. The pilot-to-production pathway for any LLM feature should include an observability gate: the system is not production-ready until the team can answer the question 'is it getting better or worse?' in under five minutes.

For platform and SRE engineers, LLM observability represents a genuine extension of the craft, not merely the application of existing tooling to a new workload. The signals, failure modes, and remediation patterns differ. Latency SLOs transfer directly. Error-budget policies need only minor adaptation. But quality SLIs, semantic drift detection, LLM-as-judge calibration, and model-incident post-mortems are new territory. Teams that treat them as new territory and invest accordingly build systems that remain mission-aligned as models, prompts, and workloads evolve. Teams that do not will repeatedly face the scenario described above.

11.3 Key Metrics: Latency, TTFT, Token Usage, and Cost

Every LLM deployment needs a core set of operational metrics before it accepts production traffic. These metrics do not answer the quality question, but they establish the baseline from which quality anomalies and cost anomalies are distinguished from infrastructure anomalies. Without a stable operational baseline, a quality regression and a GPU saturation event look identical on the dashboard.

Time-to-first-token, commonly abbreviated TTFT, is the single most important latency metric for streaming LLM responses. It measures the elapsed time from request submission to the arrival of the first token in the response stream. Users perceive TTFT as the model's response time even when the full generation runs for several additional seconds. A system with a 95th-percentile TTFT under 500 milliseconds feels responsive even if the complete response takes 4 seconds. A system with a TTFT above two seconds feels slow regardless of throughput. Every serving architecture decision described in earlier chapters, from speculative decoding to KV cache sizing, should be evaluated first for its impact on TTFT.

Total response latency, measured from request receipt to the last token, governs synchronous use cases in which the consumer waits for the complete response before acting. Tokens per second is the throughput metric: it measures how quickly the model generates once generation has begun, and it reveals GPU utilization, batch-size effects, and serving bottlenecks that TTFT alone cannot distinguish. For batch workloads, end-to-end job latency replaces TTFT as the primary latency SLI.

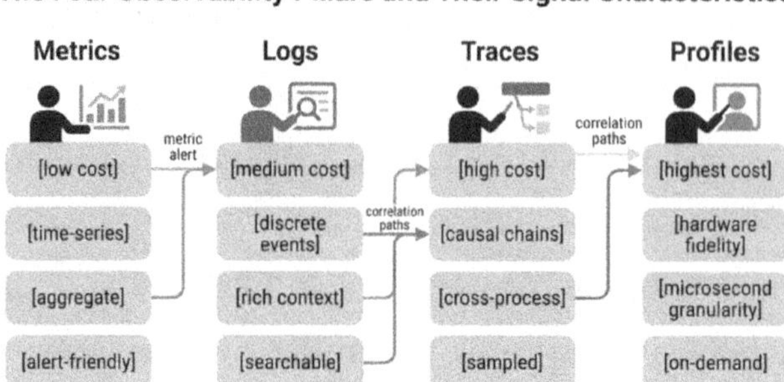

Token usage metrics decompose into prompt tokens, completion tokens, and total tokens per request. These three figures, emitted on request and aggregated by model, feature, and tenant, are the foundation of cost observability. Prompt token count reflects input complexity and prompt design; completion token count reflects output verbosity and generation strategy. Tracking the ratio of completion tokens to prompt tokens over time surfaces changes in model behavior before they manifest as cost anomalies or user complaints. A sudden increase in average completion token count often signals a prompt regression before any quality evaluator catches it.

Cost per request should be calculated in real time and attributed to the originating feature, workflow, or user segment. The formula is straightforward: tokens multiplied by the per-token price for the model tier in use, summed across prompt and completion directions. For self-hosted models, the equivalent metric is GPU compute time allocated per request, amortized against the cluster cost. Cost-per-request dashboards belong on the same screen as latency dashboards for any team that operates under a budget constraint, which is every team.

11.4 Quality Metrics: Faithfulness, Helpfulness, and Refusal Rates

Operational metrics indicate whether the system is running. Quality metrics indicate whether the system is working. The gap between those two questions is where silent regressions live. A production LLM system requires at least one quality service-level indicator in place before it can be called production-ready, and the team should not need a product manager's question to discover that the quality SLI is missing.

Faithfulness measures whether the model's response is consistent with the grounding documents or retrieved context provided to it. It is the primary quality SLI for any retrieval-augmented generation system. A response that contradicts or ignores the retrieved context is a faithfulness failure, regardless of whether it reads confidently. Faithfulness scoring can be approximated at production scale using an automated evaluator: the evaluator receives the retrieved context and the generated response and scores the overlap. The score is noisy at the individual-request level but reliable as a population-level trend over thousands of requests.

Distributed Trace Anatomy for a Training Step

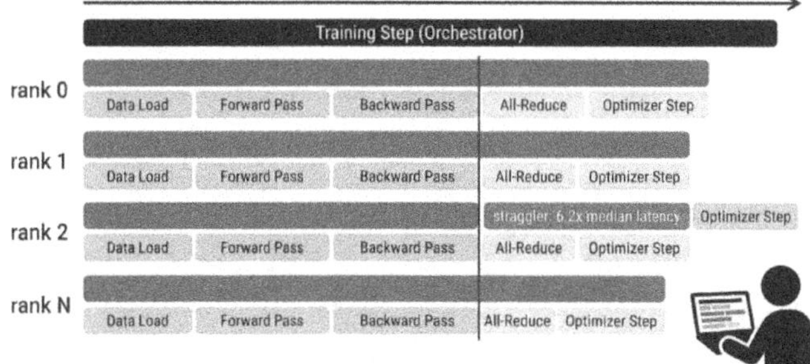

Helpfulness rubrics address the harder question of whether the response actually served the user's need. Helpfulness is inherently more subjective than faithfulness, so the rubric must be explicit and the evaluator calibrated against human ratings before it is trusted. A common approach defines helpfulness across three dimensions: task completion (did the response address the user's request), accuracy (was the information correct given available context), and presentation (was the response structured appropriately for the format). Each dimension is scored on a simple scale, and the composite score is tracked as a rolling average. When the rolling average drops below the SLI threshold, an alert fires.

Refusal rate is the fraction of requests in which the model declined to respond, either because the request triggered a safety filter or because the model judged it out of scope. Refusal rate has two failure directions. An under-refusal rate that trends toward zero may indicate that safety guardrails have eroded. An upward-trending over-refusal rate may indicate that the model is becoming overly conservative, rejecting legitimate requests and degrading the user experience. Both directions are SLI violations. Both require different remediation. Tracking them separately and setting thresholds for both is the responsible-by-design approach to safety observability.

11.5 Semantic Drift and Behavioral Drift Detection

Drift is the silent opponent of every LLM system in production. A system that performs well at launch will drift over time as the upstream model is updated, the distribution of user inputs shifts, prompt templates accumulate edits, and the retrieval corpus changes. None of these changes necessarily triggers an operational alert. All of them can degrade quality below the SLI threshold.

Drift detection is the practice of instrumenting a system to surface these changes before they reach users in volume.

Input-space drift detection monitors the statistical distribution of incoming requests over time. The canonical technique embeds each request using a fast embedding model and tracks the distribution of embedding vectors within a reference window against that of the current window. When the distance between distributions, measured by metrics such as the maximum mean discrepancy or the population stability index, exceeds a threshold, an alert fires. The alert does not indicate a quality problem directly; it indicates that the request distribution has shifted and that the quality SLIs should be monitored more closely. Input drift is often the leading indicator of quality drift by a margin of days to weeks.

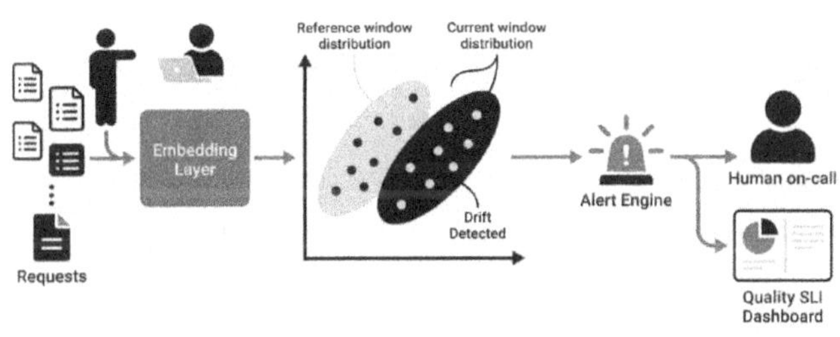

Input-space drift is often the earliest warning of quality regression.

Output-space drift detection applies a complementary approach to model responses. Embedding the response corpus and monitoring its distribution catches behavioral drift: cases where the model's output style, length, or semantic content has shifted even when the input distribution has not. Output drift often emerges after a model provider updates a backend model version without a coordinating announcement, or after a system prompt change propagates through a caching layer on a different schedule than intended. Monitoring output embeddings alongside input embeddings provides the team with two independent drift signals, each capable of catching regressions that the other might miss.

Behavioral drift detection extends beyond the embedding space to track observable behavioral properties, such as response-length distribution, refusal rate, formatting adherence (whether the model still returns JSON when asked), and citation density for RAG systems. These behavioral metrics are cheap to compute, easy to alert on, and directly interpretable by an engineer without domain expertise. A system that tracks five behavioral metrics across a rolling seven-day window will surface most prompt regressions and model update effects within hours of their introduction.

11.6 Logging Strategies for LLM Requests and Responses

Logging an LLM system presents a tension that does not exist in the same form for traditional APIs. The most valuable log content for debugging and quality analysis is the raw request-response pair: the prompt sent to the model and the completion returned. That same content often contains sensitive, regulated, or both user data. Resolving that tension requires an explicit logging policy, not an informal convention, before the system handles production traffic.

Structured logging is the foundational requirement. Every LLM request should produce a log event with a defined schema: request identifier, timestamp, model identifier, token counts (prompt and completion), latency components (TTFT and total), quality scores if available, and a status code that distinguishes model errors from infrastructure errors from safety filter activations. The schema should be versioned, and the version should appear in every log event, so that downstream consumers can handle schema evolution without breaking.

LLM Log event schema and field classification

Identity Fields	Operational Fields	Quality Fields	Redaction Status
request_id	ttft_ms	faithfulness_score	
model_id	total_latency_ms	evaluator_version	
session_id	prompt_tokens	refusal_flag	
	completion_tokens		
	cost_usd		

Retained   Redacted by default

A versioned, structured schema enables consistent analysis without manual field hunting.

Redaction policy must be set before logs are written, not after. The default posture for any field that could contain user-supplied text should be redact-by-default: the raw prompt and raw completion are excluded from logs unless an explicit retention decision has been made and documented. When retention is approved, the retained fields must be protected with the same access controls and retention schedules as the underlying user data. Logging the full conversation history of a user who later submits a deletion request creates an obligation that is expensive to fulfill. Redacting by default eliminates that obligation at the source.

Sampling strategies address the volume challenge for high-throughput LLM systems. Logging every request at full fidelity may be operationally and financially impractical beyond tens of thousands of requests per hour. A tiered sampling strategy retains one hundred percent of requests that trigger quality alerts, user feedback events, or error conditions; a configurable percentage of normal-path requests for trend analysis (typically one to ten percent); and zero percent of requests below a cost floor. The sampling rate for each tier should be configurable, not a hard-coded constant, so that the team can adjust coverage during incidents without a code deploy.

11.7 Distributed Tracing for Multi-Step Pipelines

A single user request to a modern LLM system may trigger a retrieval query, one or more reranking operations, a primary model call, a safety classifier call, and a post-processing step, all before a response reaches the user. When that sequence produces a bad response or a timeout, a log search across five separate services is the wrong debugging experience. Distributed tracing is the right one. It stitches the entire request lifecycle into a single trace that an engineer can inspect, measure, and compare across requests without cross-referencing multiple log streams.

The tracing model for LLM pipelines follows the same parent-child span structure as conventional distributed tracing. The root span covers the end-to-end request lifecycle. Child spans are created for each stage: retrieval, embedding, model inference, tool invocation, and post-processing. Each span carries the latency of that stage, the input and output token counts where applicable, and any error conditions. The trace is propagated via a trace context header that flows through every service call in the pipeline, including calls to external model APIs if the provider supports header passthrough.

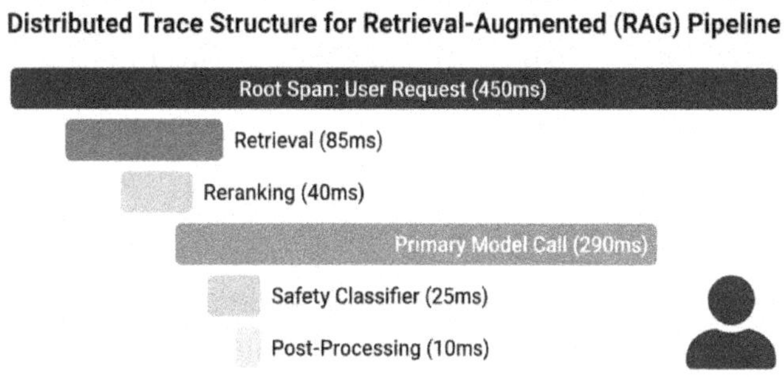

Distributed Trace Structure for Retrieval-Augmented (RAG) Pipeline

Root Span: User Request (450ms)
Retrieval (85ms)
Reranking (40ms)
Primary Model Call (290ms)
Safety Classifier (25ms)
Post-Processing (10ms)

Distributed tracing is the only practical way to debug latency and qualityissues in multi-step LLM pipelines.

For agentic pipelines, where the number of model calls and tool invocations is not known at the start of the request, the tracing model must accommodate dynamic span creation. The trace context should be passed through the agent loop as a first-class parameter, and each iteration of the loop should create a new child span tagged with the iteration index. Tool invocations within the loop should create grandchild spans that capture the tool name, input parameters (subject to redaction policy), and result status. A completed agentic trace should allow an engineer to reconstruct the entire reasoning sequence from a single trace view.

The trace retention policy follows the same logic as the log retention policy. Traces containing user-supplied content should be governed by the same redaction rules as structured logs. Traces for error and quality alert paths should be retained at 100% for at least the incident review window. Traces for normal-path requests can be sampled and aggregated into performance histograms after a shorter retention period. A trace that captures per-span latency but redacts per-span content is sufficient for most debugging workflows and eliminates most regulatory exposure.

11.8 User Feedback Loops and Quality Signals

User feedback is one of the highest-signal quality sources available to an LLM system and one of the most consistently underutilized. The pattern seen in the opening scenario, a feedback widget that writes to a database no one monitors, is common enough to be considered the default state rather than an exception. Closing that gap requires treating user feedback as a quality signal with the same operational discipline as a latency metric: it must be ingested, aggregated, alerted on, and visible to the same engineers who own the rest of the stack.

Explicit feedback signals include thumbs-up and thumbs-down widgets, star ratings, correction submissions, and free-text feedback forms. These signals are high-value but low-volume. A thumbs-down event represents a user who cared enough to interact with the feedback widget, which means it almost certainly reflects a genuine quality failure. Explicit feedback should be captured with the associated request identifier so it can be joined to the structured log event and the distributed trace for that request. A feedback event without a linked request is an opinion; a feedback event with a linked request is a debugging artifact.

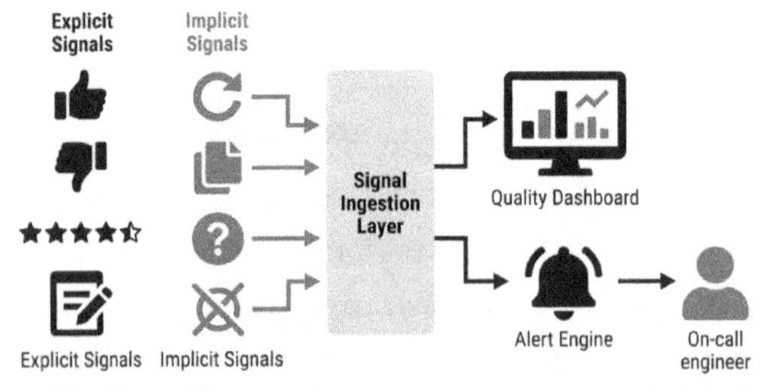

Feedback signals belong on the same dashboard as latency metrics, not in a separate database no one checks.

Implicit feedback signals are higher-volume and lower-noise than most teams expect. A user who clicks the regenerate button is implicitly signaling that the previous response was insufficient. A user who copies a response to the clipboard is implicitly signaling that the response was useful. A user who asks a follow-up question that rephrases the original request is implicitly signaling that the first response did not fully address the need. A user who abandons a session within 30 seconds of receiving a response implicitly signals dissatisfaction. Each of these behavioral patterns can be instrumented without explicit user action or personally identifiable information.

Wiring feedback signals into the shared observability stack requires two engineering decisions: where the signals are ingested and how they are aggregated. Ingestion should happen at the application layer, not the client layer, to ensure that the signal is captured regardless of client-side ad blockers or network interruptions. Aggregation should produce both per-request flags (was this specific response regenerated?) and rolling population metrics (what fraction of responses in the last hour were regenerated?). The population metric is the SLI. The per-request flag is the debugging anchor when the SLI breaches its threshold.

11.9 Real-Time Evaluators and LLM-as-Judge in Production

User feedback provides a quality signal at the rate users choose to give it. Real-time evaluators provide a quality signal at the rate the system generates responses. For high-volume systems where user feedback is sparse relative to request volume, automated evaluators are the only practical mechanism for continuous quality monitoring. The dominant pattern is LLM-as-judge: a secondary model, typically smaller and cheaper than the primary, scores each production response against a rubric defined by the team.

Rubric design is the most consequential decision in deploying an LLM-as-judge evaluator. A rubric that is too coarse (score from one to five with no criteria) produces scores that are internally consistent but uninformative for root-cause analysis. A rubric that is too granular (a thirty-item checklist) produces scores that are expensive to generate and difficult to aggregate. The recommended approach defines three to five criteria that correspond directly to the quality dimensions that matter for the specific use case: faithfulness, task completion, format adherence, and tone are common choices for customer-facing assistants. Each criterion is scored on a three-point scale (meets standard, partially meets, does not meet) with an explicit definition for each level.

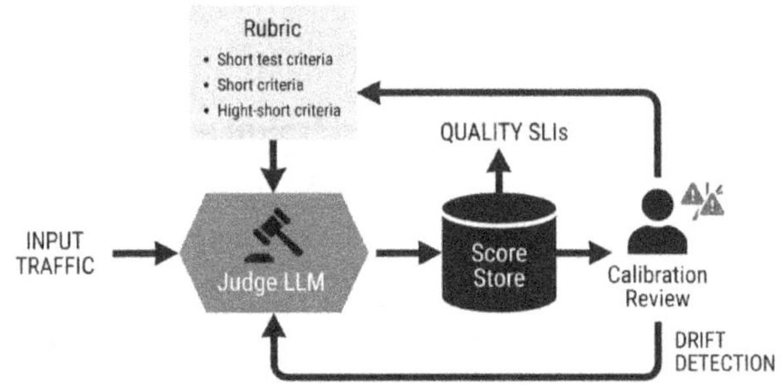

An LLM-as-judge pipeline requires calibration maintenance or it becomes a source of false confidence.

Calibration is the most commonly skipped step and the one most likely to undermine the value of the entire evaluator. Calibration involves comparing evaluator scores against human ratings on a held-out sample and measuring the agreement rate. An evaluator that agrees with human raters at 75% or higher is sufficiently reliable for trend-level quality monitoring. An evaluator who agrees at sixty percent or below is producing scores that are less informative than a random sampling of user feedback. Calibration should be run at launch, after any change to the evaluator model or rubric, and on a scheduled basis (monthly is a reasonable default). If calibration agreement degrades below the threshold, the evaluator should be flagged as unreliable, and the team should not use its scores to close an error-budget burn alert.

Evaluator drift is a second-order problem: the evaluator model can drift independently of the primary model, creating the appearance of stable quality when the primary model has regressed, or vice versa. Guarding against evaluator drift requires a fixed evaluation dataset, drawn from historical production traffic and scored by human raters, against which the evaluator is benchmarked on each calibration cycle. If the evaluator's scores on that fixed dataset shift between cycles, the evaluator has drifted. The fix is to retrain or replace the evaluator, not to adjust the quality SLI threshold.

11.10 Alerting, SLOs, and SLA Design for LLM Systems

Service-level objectives for LLM systems require more dimensions than the latency and availability SLOs that govern most API services. A complete LLM SLO set covers four distinct signal classes: operational signals (latency, availability, error rate), quality signals (faithfulness, helpfulness, refusal rate), cost signals (cost per request, token usage), and freshness signals (how recently the retrieval corpus or model weights were updated). Each signal class may require different measurement windows, different breach response procedures, and different stakeholder notifications.

Per-tier SLO design recognizes that different user segments or use cases warrant different guarantees. An internal productivity assistant and a customer-facing support bot may share the same underlying model but require different SLO targets because the consequences of a quality breach differ. The internal assistant may tolerate a quality SLI below threshold for a four-hour error-budget burn window before escalation; the customer-facing bot may require a one-hour window. Encoding these distinctions at the SLO level, rather than the alert rule level, gives the on-call engineer a clear framework for prioritizing incidents without needing context about business impact in the middle of a page.

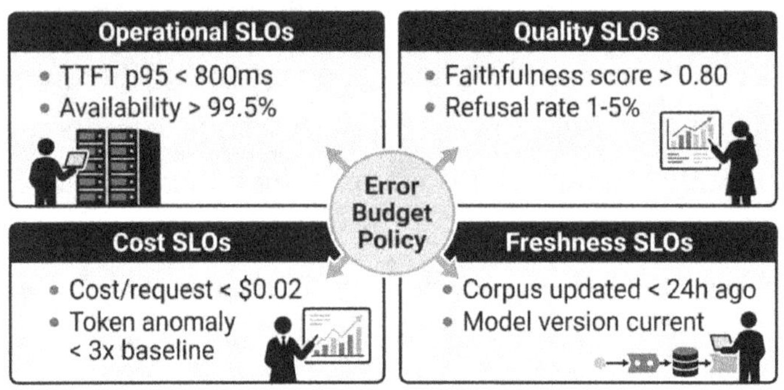

Four-dimensional SLO design gives the team a complete and actionable production health definition.

Error-budget policies for LLM systems follow the same burn-rate algebra as conventional SRE practice. The new consideration is that quality SLIs produce burn events that do not correlate with operational SLI burns. A quality regression can consume quality error budget while the operational budget remains untouched, and vice versa. The release gate policy should require that both budgets are above the threshold before a deployment is approved. A deployment that improves latency but degrades quality is not successful.

Alerting thresholds for quality SLIs require a burn-in period after deployment to establish a stable baseline. Setting a quality SLI threshold on day one of production, before the evaluator has scored enough requests to produce a reliable distribution, results in either excessive false positives or excessive false negatives, depending on which direction the threshold was set. The recommended approach runs the evaluator in observe-only mode for the first two weeks of production, uses the resulting distribution to set thresholds at the fifth percentile of historical scores, and activates the alert on day fifteen.

11.11 Cost Observability and Anomaly Detection

Cost observability for LLM systems differs from cost monitoring for conventional compute workloads in one important way: costs are generated per token, which means a single pathological request or a single prompt regression can drive cost anomalies that are both large in magnitude and invisible on a request-count dashboard. A batch job that inadvertently sends 10,000-token prompts instead of 1,000-token prompts will appear as normal traffic by every operational metric until the billing statement arrives, unless cost-per-request is tracked in real time.

Cost dashboards should present at least three views. The first is cost per request by model and feature, trended over a rolling window. This view surfaces the per-request cost baseline and makes anomalies visible against a stable reference. The second is token usage decomposition: prompt tokens versus completion tokens, trended separately. A sudden increase in prompt token count often indicates a context injection error or a caching miss; a sudden increase in completion token count often indicates a generation verbosity regression. The third view is cumulative cost against budget, broken down by feature and time period. This view belongs in the product and engineering leadership dashboard, not only the engineering one.

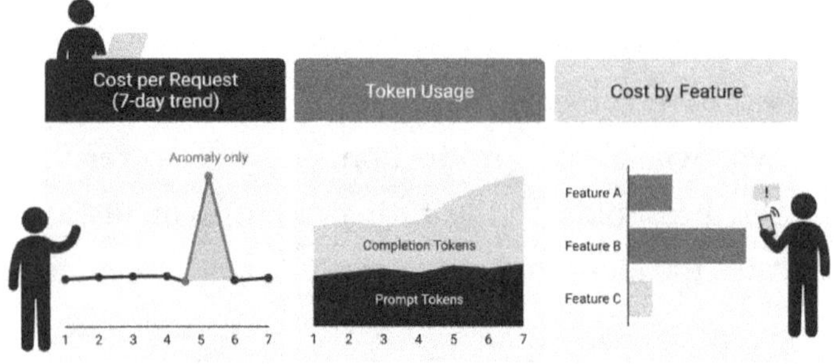

Cost dashboards that decompose by feature and token direction catch regressions before they become budget incidents.

Anomaly detection for cost requires a reference model that accounts for normal variation in request volume and prompt complexity. A simple threshold alert on hourly cost will produce false positives during expected traffic spikes and miss regressions that occur during low-traffic periods. A more reliable approach calculates cost-per-request and applies a statistical anomaly detector to that per-request value, isolating cost anomalies from volume anomalies. When the per-request cost exceeds three standard deviations of the rolling baseline, the alert fires regardless of whether the total hourly cost has breached any absolute threshold.

Budget attribution connects cost observability to financial governance. Each feature or product surface that calls the LLM should carry a cost center tag that flows through every token-usage log event. Monthly cost roll-ups by cost center enable finance and engineering leadership to assess ROI at the feature level, not just at the platform level. A feature that generates 10% of the value at 40% of the LLM cost is a candidate for model downgrade, caching, or an architectural rethink. That analysis is only possible when attribution is in place.

11.12 Postmortems and Root-Cause Analysis for Model Incidents

LLM incidents resist the root-cause analysis patterns that work well for conventional software failures. A conventional incident has a clear causal chain: a configuration change caused a service to fail, the failure propagated to dependent services, and the timeline reconstruction is deterministic. An LLM incident may involve a model whose outputs are non-deterministic, a quality regression that emerged gradually over weeks, a contributing cause from an upstream model update that was not announced, and a detection gap because the quality SLI was not instrumented. Applying a conventional post-mortem template to that incident produces a document that is technically complete but operationally useless.

LLM-specific post-mortem templates add four categories that conventional templates do not include. The first is model version context: which model version was in use at the time of the incident, and whether a model update occurred within the detection window. Many LLM quality regressions trace back to a model provider update that changed output behavior in ways that were not tested. The second is prompt version context: which prompt template was in use, and whether any prompt changes were deployed in the period before the incident. Prompt changes are deployments and should be tracked in the same change log as code deployments.

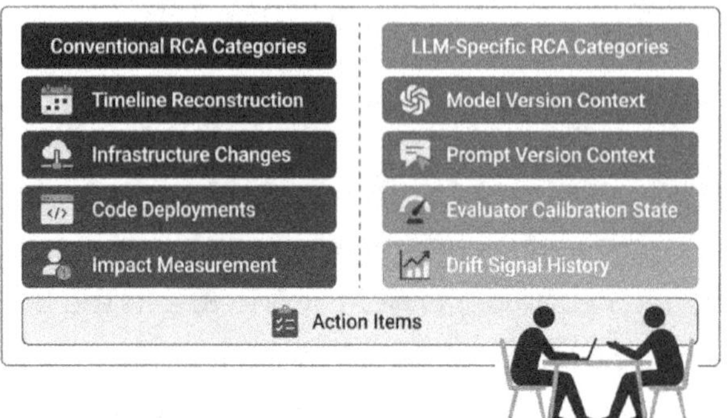

LLM incidents require four additional root-cause categories that conventional templates omit.

The third LLM-specific category is evaluator calibration state: at the time of the incident, was the automated evaluator calibrated and within its agreement threshold? If the evaluator had drifted, the quality SLI may have been reporting false confidence, masking the regression. Documenting the evaluator's state at the time of the incident is essential for understanding why the detection gap existed. The fourth category is drift signal history: were there drift alerts or near-misses in the weeks before the incident that should have triggered a quality review? Drift signals that fired and were dismissed without action are a systemic problem that action items from the post-mortem should address directly.

Timeline reconstruction for LLM incidents requires logging discipline that many teams do not yet have in place. A useful incident timeline needs at minimum: the timestamp and content of every prompt change deployed in the preceding thirty days, the timestamp of any model version changes (from the model provider's change log, where available), the history of quality SLI scores and drift alerts for the preceding thirty days, and the first user feedback event that could have signaled the regression. Building that timeline after the fact, without pre-instrumented logging, is a multi-day exercise. Building it from well-structured logs and traces takes under an hour.

11.13 An Observability Maturity Model for LLM Platforms

A maturity model gives a team an honest baseline and a target. It converts an abstract capability like 'good observability' into a concrete set of stages that can be assessed, planned against, and progressed through. The five-stage model below describes the observability states that most production LLM platforms occupy, ordered from least to most capable. A team that can honestly place itself on this model has the information it needs to prioritize the next investment.

Stage one is ad hoc logging. The team has no structured log schema for LLM requests. Quality assessment happens through manual spot-checking. Cost is monitored only at the monthly billing level. User feedback is collected but not analyzed. Drift is not detected; user complaints discover regressions. Most teams are at stage one at launch. Some remain there indefinitely.

Stage two is operational instrumentation. The team has structured logs with a defined schema for latency, token usage, and error classification. TTFT and cost-per-request are tracked and alerted on. Distributed tracing covers at least the primary model call. User feedback is ingested but may not be wired to a shared dashboard. No quality SLI is in production. Stage two is the state most teams reach within the first three months of a serious production deployment.

LLM Observability Maturity Model

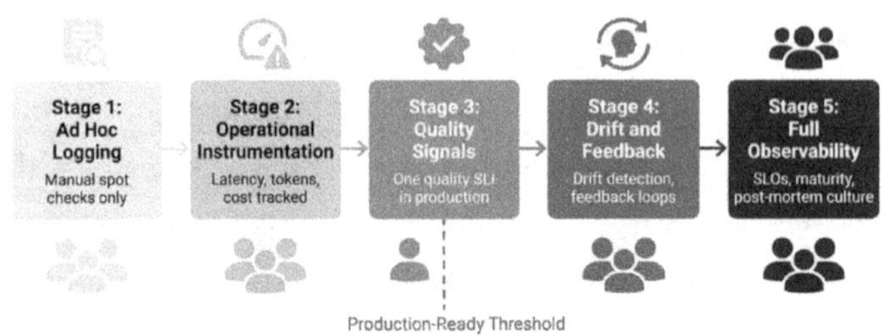

Most teams launch at Stage 1. Production-ready requires reaching at least Stage 3.

Stage three is quality signal coverage. The team has at least one quality SLI in production, visible on a shared dashboard. An automated evaluator runs on production traffic, and its scores are aggregated into the SLI. User feedback is wired to the same alert channel as operational metrics. Stage three is the production-ready threshold for this chapter: a system that cannot answer the quality question is not production-ready, regardless of its operational health.

Stage four is drift and feedback integration. Input-space and output-space drift detection are in place. User feedback signals (explicit and implicit) feed the same quality dashboard as the automated evaluator. The evaluator is calibrated on a scheduled basis, and its calibration state is tracked. Prompt changes are logged as versioned deployments and appear in the incident change log. Stage four is the target state for any system that has been in production for more than six months.

Stage five is full observability. The team has SLOs for all four signal classes (operational, quality, cost, and freshness) and error-budget policies that govern release decisions. Post-mortems follow an LLM-specific template, and action items are tracked to closure. Drift signals have a documented response playbook. The team can answer the question 'is it getting better or worse?' in under five minutes, for any feature, at any time. Stage five teams spend less time debugging incidents and more time improving quality, because the observability layer closes the feedback loop automatically.

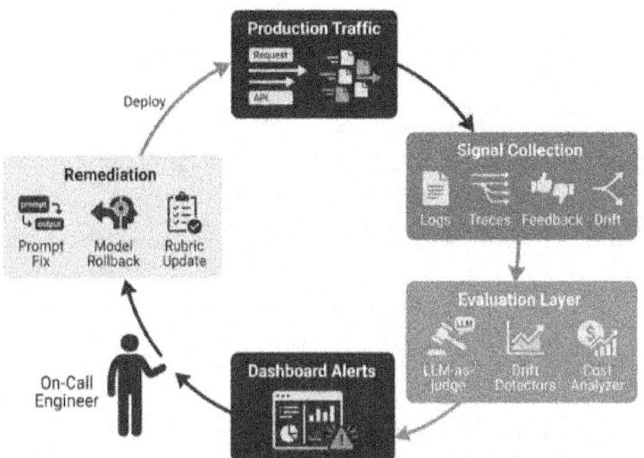

Full observability closes the feedback loop from production signals to remediation automatically.

11.14 Technical Checklist

Use this checklist as a release gate before declaring an LLM system production-ready against the observability requirements of this chapter.

1. Operational metrics include time-to-first-token, tokens-per-second, total response latency, cost-per-request, and a status code taxonomy that distinguishes model errors from infrastructure errors from safety filter activations.
2. Token usage is decomposed into prompt tokens and completion tokens, and each is attributed to the originating feature or cost center in every log event.
3. At least one quality SLI is in production and visible on a shared dashboard accessible to on-call engineers, not only to the team that built the evaluator.

4. The automated evaluator has been calibrated against human ratings, and the calibration agreement rate meets or exceeds the threshold set at deployment.
5. Evaluator calibration is scheduled and tracked; a failed calibration cycle triggers an alert and places the evaluator in observe-only mode until recalibration is complete.
6. Input-space and output-space drift detection are both in place for production traffic and are wired to the same alert channel as operational metrics.
7. Logs are structured with a versioned schema, redacted by default for user-supplied content, and governed by a documented retention policy by field class.
8. Distributed traces span retrieval, model calls, tool invocations, and post-processing stages, with trace context propagated through every service call in the pipeline.
9. User feedback signals (thumbs-up, thumbs-down, regenerate, copy, session abandonment) are ingested at the application layer, linked to request identifiers, and wired to the shared quality dashboard.
10. SLOs are defined across all four signal classes: operational, quality, cost, and freshness. Error-budget policies are documented and applied in release-gate decisions.
11. Post-mortems for LLM incidents follow a template that includes the four LLM-specific root-cause categories: model version context, prompt version context, evaluator calibration state, and drift signal history.
12. The team can place itself on the five-stage observability maturity model and has a documented plan to reach the next stage within a defined time period.
13. Cost anomaly detection uses per-request cost against a rolling statistical baseline, not only an absolute monthly threshold.

14. Prompt changes are logged as versioned deployments and appear in the same change log as code deployments, enabling incident timeline reconstruction without manual artifact hunting.

11.15 Team Conversation

The following questions are designed for technical leads to use in team meetings to pressure-test alignment on observability practices before and after a production deployment.

1. Could the team answer the question 'is it getting better or worse?' for the most important LLM feature today, using existing dashboards, in under five minutes? If not, which signal class is missing, and what is the plan to add it?
2. Which quality SLI would have caught the last quality regression the team experienced? Was that SLI in place at the time, and if not, what prevented it from being instrumented?
3. Where does drift detection currently stand? Are both input-space and output-space drift detectors in place and alerting to the same channel as latency pages?
4. Are user feedback signals visible to the same engineers who own latency? Or are they collected into a separate system that only the product team monitors?
5. When was the LLM-as-judge evaluator last calibrated? What was the agreement rate, and is that rate above the threshold the team set at deployment?
6. Do the SLOs for the primary LLM feature include quality and freshness, or only latency and availability? If quality and freshness are absent, what would it take to add them in the next sprint?

7. Where does the team honestly sit on the five-stage observability maturity model? What is the single highest-leverage investment to move to the next stage?
8. Has the post-mortem template for LLM incidents been updated to include model version context, prompt version context, evaluator calibration state, and drift signal history? If not, which of those categories would have changed the action items from the last incident?
9. What is the current cost-per-request for the primary LLM feature, and is that figure tracked in real time on a shared dashboard? Is there an anomaly detector that would catch a sudden cost regression independent of volume?
10. If the upstream model provider released a silent model update tonight, how long would it take the team to detect any resulting quality regression? What is the detection pathway?

11.16 Key Takeaway

LLM observability is a distinct engineering discipline, not a variant of conventional API monitoring. Latency and error rate are necessary starting points, but they answer a different question than the one that matters most: whether the system produces useful, accurate, on-policy responses. Answering that question requires high-quality SLIs, automated evaluators, and user-feedback signals instrumented with the same operational discipline as latency metrics. Teams that treat quality as an afterthought, measured only when a stakeholder asks, will consistently discover regressions through user complaints rather than through their observability stack. The goal of this chapter's playbook is to make that situation structurally impossible.

Drift detection and feedback loops are the mechanisms that keep a production LLM system aligned over time. Models change.

Prompts drift. User populations evolve. Retrieval corpora go stale. Any of these changes can degrade quality without touching the operational dashboard. Input-space drift detection, output-space drift detection, and behavioral metric tracking together provide a three-signal early warning system that surfaces regressions before they accumulate into incidents. User feedback, wired correctly, turns every thumbs-down and every regenerate click into a real-time quality signal that the on-call engineer sees alongside the latency histogram. Closing these feedback loops is the mission-aligned path to operational clarity over a non-deterministic system.

The five-stage observability maturity model gives every team an honest self-assessment framework and a clear improvement path. Most teams launch at stage one. Stage three is the production-ready threshold: at a minimum, one quality SLI must be in production, visible, and alerting before a system can be called production-ready against this chapter's standard. Stage five teams close the full feedback loop from production signals through automated evaluation through remediation, and they can answer the quality question at any time without a two-week investigation. The investment to move from stage one to stage three is modest relative to the cost of the incidents it prevents. The investment to reach stage five is the cost of operating a system the team actually trusts.

12 Testing, CI/CD, and Model Validation

12.1 Opening Scenario

The release looked clean on paper. The team had run the same evaluation suite they trusted for six months: thirty-seven categories, automated scoring, and a passing threshold that had never been violated in production. The model upgrade shipped on a Thursday, monitoring dashboards stayed green, and the on-call engineer closed the deploy ticket. Two days later, the support queue filled with tickets from the platform's largest enterprise customer. The complaint was consistent: answers to their most common question type had become shorter, vaguer, and less actionable. The word customers used most was 'worse.'

The incident commander pulled the eval results. They were still green. The team ran the suite again against the new model and again against the old one; both passed with nearly identical scores. The problem was not that the eval was wrong, exactly. It was that the eval had never been calibrated to production traffic shape. The enterprise customer's traffic skewed heavily toward two question categories, which the eval suite weighted at roughly 2% of total coverage. When the model update traded depth in those two categories for improvements across the other thirty-five, the eval score went up, the deployment gate opened, and the regression shipped silently to the customer who needed those two categories most.

Rollback took four hours because the release process had not standardized the artifact manifest format, and the prior model version had to be located manually across three storage locations. A shadow deployment had been discussed at the planning meeting for this release cycle, but was deferred to optional scope.

Post-incident review identified four structural gaps: eval weighting not tied to production traffic, no traffic-replay tests in the pre-production gate, no shadow comparison before the full cutover, and no reproducible artifact manifest linking model version to prompt version to eval results. Every one of those gaps had a known pattern. The team had not yet built them into the release process. This chapter is the discipline they adopted going forward.

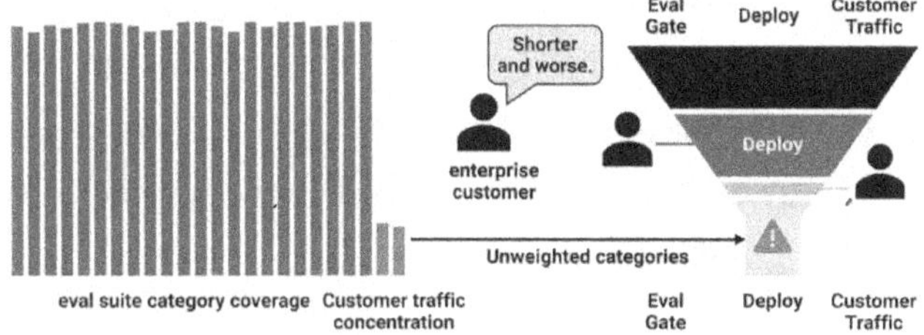

12.2 Why It Matters

Every engineering organization eventually discovers that the reliability culture it built for deterministic software does not transfer cleanly to LLM-powered systems. Unit tests pass because a function returns a string, not because the string is useful. Integration tests pass because the pipeline completes, not because the output is accurate. Performance tests pass because latency is within bounds, not because the model's reasoning is coherent. The gap between 'the system works' and 'the system works well' is large in LLM applications, and classical CI/CD pipelines were not designed to close it.

The workflow-level impact of this gap is significant. When a regression escapes the pre-production gate and reaches production, the cost is not just engineering time. It is customer trust that is eroded by degraded answers, support tickets that take human time to resolve, and, in regulated domains, audit exposure from outputs that the team cannot explain or trace. The eval suite that fails to catch a quality regression is not a neutral artifact; it is a false signal that actively accelerates the release of a degraded product. Building a mission-aligned testing discipline means treating quality gates with the same engineering rigor applied to uptime and latency.

The patterns in this chapter form a coherent system. Unit tests for prompt logic and orchestration, eval suites calibrated to production traffic, integration tests across pipeline components, traffic-replay harnesses, shadow deployments for pre-cutover comparison, and continuous evaluation wired to the same alerting paths as operational metrics: each layer closes a different failure mode. The CI/CD checklist at the end of the chapter is the structural difference between a team that ships LLM changes with confidence and one that ships them with anxiety.

> *This chapter and Chapter 4 are two halves of one discipline, and this is the test-patterns half. Chapter 9 owns the concrete machinery of verifying prompt and model behavior: the unit tests, the eval suites with golden sets, the LLM-as-judge patterns, and the continuous-integration gates that catch a regression before it reaches production. The lifecycle and governance frame those tests run inside is Chapter 4's territory. A reader who needs the prompt versioning policy or the prompt review workflow should turn to Chapter 4, which*

establishes the lifecycle this chapter's tests operate within.

12.3 Unit Tests for Prompts and Model Outputs

Unit testing an LLM application requires a deliberate split between the parts of the system that are deterministic and the parts that are not. The orchestration layer, the input transformation pipeline, the retrieval query construction, the output parsing logic, and the guardrail routing rules: all of these can be tested with classical unit tests because they are functions that take inputs and produce outputs without calling the model. Testing them in isolation quickly and cheaply catches the largest category of regressions, and it keeps the test suite fast by avoiding model calls on every commit.

Model outputs themselves require a different assertion style. The primary pattern is soft assertion: instead of asserting exact string equality, the test asserts that the output satisfies a predicate. The predicate might check that the output contains required keywords, falls within a length range, matches a structural schema (valid JSON, correct field names), avoids prohibited phrases, or passes a classifier. Soft assertions tolerate the natural variation in model outputs without becoming flaky, and they document the intended contract between the system and the model in executable form.

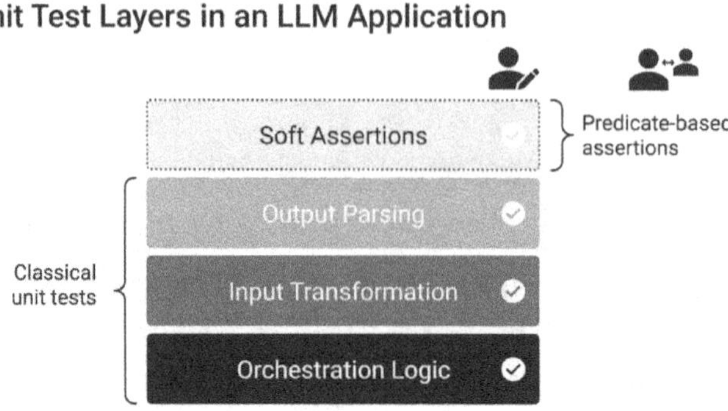

Mock-mode testing is the third pattern. In mock mode, the orchestration layer is wired to a stub that returns pre-recorded model responses instead of calling the live API. Mock mode tests verify that the orchestration logic handles the model's response correctly: it routes a refusal to the fallback handler, parses a structured output and propagates its fields, retries on transient errors, and respects the retry budget. These tests are fast, deterministic, and safe to run on every commit because they do not consume model API credits or depend on model availability.

The fourth pattern is snapshot testing for prompt templates. A snapshot test captures the rendered prompt for a given set of inputs and fails if the rendered output changes unexpectedly. Snapshot tests catch the category of regression where a template refactoring accidentally removes a critical instruction, rearranges context in a way that degrades model performance, or introduces a formatting artifact. The snapshot file lives in version control alongside the template, so every change to the template requires a deliberate review of the rendered output. This pattern is especially valuable in systems with complex multi-turn prompt construction, where small rendering changes have large effects on model behavior.

12.4 Eval Suites: Golden Sets, Rubrics, and LLM-as-Judge

An evaluation suite is the team's formal definition of what 'good' means for the system. Building a suite that deserves trust requires answering three questions before writing a single test case: what behaviors does the system need to exhibit, how will those behaviors be measured, and how will the weighting of test cases reflect the actual distribution of production traffic. Suites that skip the third question are the most common source of the false-green eval described in the opening scenario.

Golden sets are the foundation of most eval suites. A golden set is a curated collection of input-output pairs in which the expected output represents the correct or ideal response to that input. The outputs are authored by subject-matter experts, collected from highly rated production traffic, or synthesized from domain documentation. Golden sets work best for closed-domain tasks where correct answers can be specified precisely, such as information extraction, classification, summarization with required facts, structured data generation, and code generation with testable correctness criteria. For open-ended tasks with no single correct answer, golden sets need to be paired with rubrics.

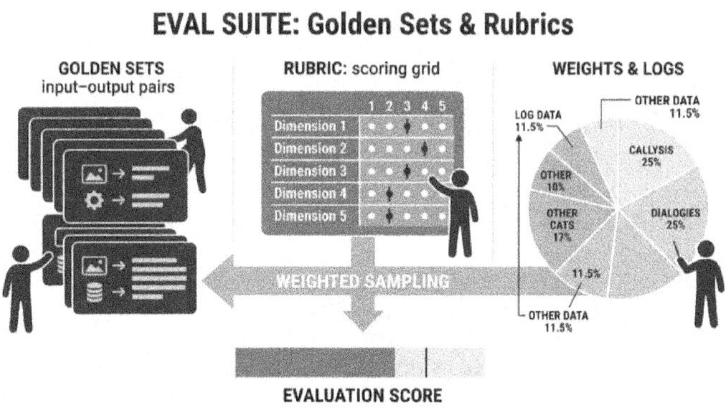

Rubric-based scoring defines the dimensions of quality explicitly and scores each output against each dimension on a numeric scale. A rubric for a customer-service response might score accuracy, completeness, tone appropriateness, conciseness, and avoidance of prohibited content separately, then combine the dimension scores with weights that reflect what matters most for that task. Rubrics make the definition of quality transparent, reviewable, and adjustable. When the team disagrees about whether a model improvement is real, a rubric gives them a shared vocabulary for the argument.

LLM-as-judge is the pattern that scales rubric-based evaluation beyond what human reviewers can cover at release cadence. A judge model receives the rubric, the input, and the evaluated output, and returns a score with a rationale. The pattern dramatically increases eval throughput, but it introduces a new failure mode: the judge model has its own biases, verbosity preferences, and blind spots. A judge model that has not been calibrated against human review is a proxy metric, not a ground truth, and treating it as a gate without calibration reproduces the false-confidence problem at a different layer.

Calibration is what distinguishes a trustworthy LLM-as-judge from a convenient fiction. Calibration runs a representative sample of outputs through both the judge model and human reviewers, measures the correlation between their scores on each rubric dimension, and identifies the dimensions where the judge diverges most from human judgment. Dimensions with low correlation get either downweighted in the overall score or replaced with a human-review path. The calibration should be documented, versioned alongside the judge prompt, and refreshed on a scheduled cadence because both the judge model and the system under evaluation drift over time. A cadence of four to eight weeks is common in active production environments.

12.5 Integration Testing Strategies for LLM Pipelines

Unit tests verify individual components in isolation. Integration tests verify that the components work correctly when connected. In an LLM pipeline, the integration surface is large: the retrieval system that provides context, the prompt construction layer that assembles that context into a model input, the model itself, the output parser that structures the response, the guardrail layer that filters or routes it, and the application layer that presents it to the user. Any one of these boundaries can introduce a regression that unit tests miss entirely.

End-to-end integration tests exercise the full path from a user input to a system output using a real or near-real configuration of all components. The test suite typically operates in a staging environment that mirrors production configuration, uses a production-representative set of test inputs, and asserts on output quality using the same soft-assertion patterns as the unit-test layer. End-to-end tests are slower and more expensive than unit tests, which means they run less frequently; a common pattern is to run the full suite on every pull request to the main branch rather than on every commit.

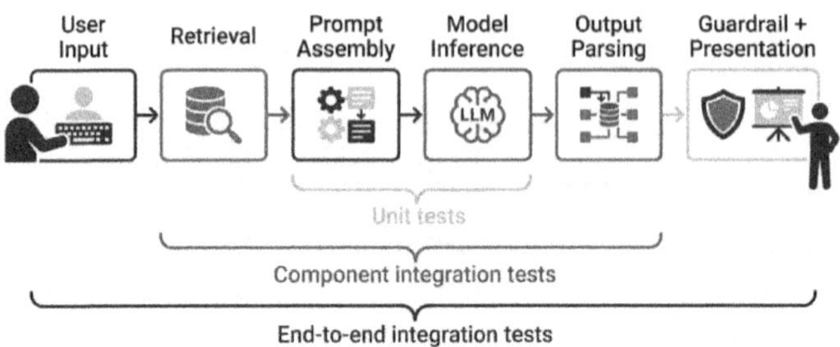

Contract testing addresses the integration points between components that are owned by different teams or deployed independently. A contract test verifies that a component honors the interface contract agreed with its consumers: that the retrieval system returns documents in the schema the prompt assembler expects, that the model API returns responses in the format the parser expects, and that the guardrail layer passes the fields the application layer depends on. Contract tests run fast because they test the interface, not the full behavior, and they catch the breaking changes that integration tests catch only after a full pipeline run.

External API dependencies require a separate testing strategy because they cannot be controlled or versioned by the team. The standard pattern is a test double hierarchy: unit tests use in-memory stubs that return pre-recorded responses; component integration tests use a record-and-replay harness that captures live API responses once and replays them on subsequent runs; end-to-end tests in staging use the live API against a rate-limited test account. The record-and-replay approach provides realism without incurring API costs on every test run and creates a corpus of real API responses that can be inspected when behavior changes.

Schema validation is an underused integration-testing technique for LLM pipelines. If the system is designed to return structured outputs (JSON responses, classified intents, extracted entities), then every integration test should include a schema validation step that asserts the output conforms to the expected structure. Schema failures constitute a different category of regression than quality failures: a model that has been updated or a prompt that has been modified may change the output structure in ways that break downstream consumers, even when content quality remains unchanged. Catching structural regressions in the integration gate rather than in production is a low-cost safety layer with high leverage.

12.6 Traffic-Replay Testing and Production-Shape Evaluation

The most reliable proxy for production behavior is production traffic itself. Synthetic test suites, however carefully constructed, reflect the team's beliefs about what users ask and how they ask it. Real traffic reflects what users actually do, including edge cases, ambiguous phrasings, long conversation threads, and domain-specific vocabulary that no one anticipated at the time of suite construction. Traffic-replay testing closes the gap between the suite and reality by using anonymized production traffic as the test input corpus.

The traffic capture process must address privacy before any replay infrastructure is built. Inputs that contain personally identifiable information, financial details, health information, or other sensitive content must be anonymized or pseudonymized before they enter the replay corpus. The anonymization approach should be documented, reviewed by the privacy team, and applied consistently. Skipping this step either prevents traffic-replay testing from being used at all (if the organization correctly refuses to process raw PII in a testing pipeline) or creates a compliance exposure that will surface during an audit.

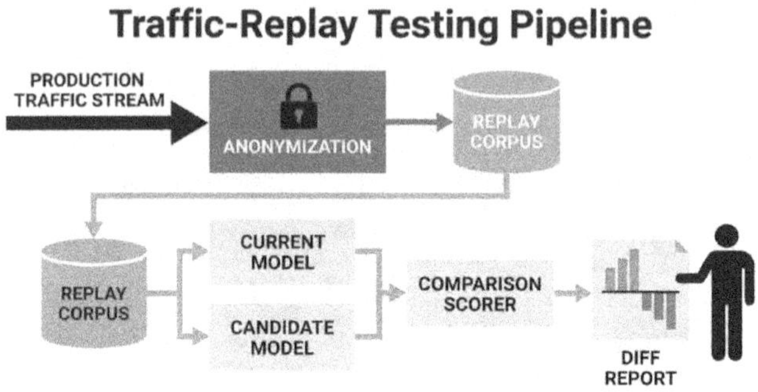

Traffic-Replay Testing Pipeline

The replay harness runs the anonymized corpus through both the current production model and the candidate model, scores both sets of outputs using the eval suite, and produces a comparison report that shows where the candidate improves, regresses, or is statistically indistinguishable from the current model. The comparison should be segmented by the same traffic categories used in the eval suite weighting, so regressions in high-traffic categories are visible even when overall scores are flat or improving.

Weighting the replay results to match production traffic shape is the step that gives traffic-replay testing its principal advantage over static eval suites. If ten percent of production traffic falls into a category where the candidate model regresses, that regression should account for ten percent of the weighted score, not two percent because the eval suite happened to include fewer examples from that category. The weighting calculation requires maintaining a traffic distribution profile that is updated regularly, typically weekly or with each major deployment, to reflect changes in user behavior over time.

Statistical significance thresholds prevent the replay harness from raising false alarms on small corpora. A regression that affects three out of fifty inputs in a low-traffic category may represent noise rather than signal. Setting a minimum sample size per category before a regression is flagged, and applying a confidence interval to the score difference, reduces alert fatigue and helps the team focus review effort on changes that are large enough to matter. The threshold settings should be documented and revisited when the corpus size changes significantly.

12.7 Canary Releases and Blue-Green Deployments

Canary and blue-green deployments are the deployment patterns that limit the blast radius of a quality regression in production. Both patterns are standard in deterministic service deployment, but they require adaptation for LLM applications because the failure signal is quality degradation rather than error rate or latency. The rollout metrics that determine whether a canary graduates must include quality signals, not just operational signals, or the pattern provides false safety.

A canary release routes a small percentage of production traffic (typically one to ten percent) to the new model version while the remainder continues to use the current version. Quality metrics are measured on both the canary cohort and the control cohort, and the difference is monitored against a threshold for a dwell period before the release graduates to a wider rollout. The canary pattern is well-suited to cases where the team wants production-representative feedback before full commitment but is willing to accept some user exposure to a potential regression.

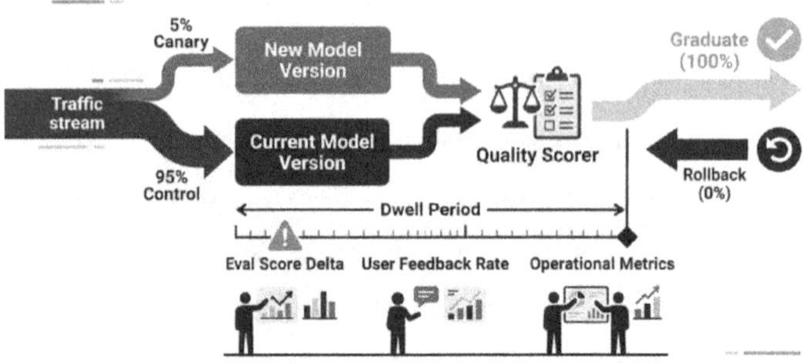

Blue-green deployment maintains two complete, production-ready environments: the current version (blue) and the new version (green). Traffic shifts from blue to green at a specific cutover point, typically controlled by a load balancer or traffic manager. If problems are detected after cutover, traffic shifts back to blue in seconds without rebuilding or redeploying. The blue-green pattern prioritizes rollback speed over gradual validation; it is appropriate for changes where the team has high confidence in pre-production testing and wants to minimize the duration of any split-traffic state.

The quality guardrail layer in a canary deployment is the part most commonly neglected. A team that monitors only latency and error rate during a canary dwell period is monitoring the wrong signals for an LLM application. The guardrail layer should continuously sample canary outputs, apply the eval rubric to that sample using an automated scorer, and compare the running score against the control cohort's running score. If the canary score falls below a defined threshold relative to control, the release automation pauses the rollout and alerts the on-call engineer before the regression reaches more users.

Rollout velocity controls the pace at which a canary graduates from the initial small cohort to full production. A staged velocity schedule might move from five percent to twenty percent after twenty-four hours, from twenty to fifty percent after another twenty-four hours, and from fifty to one hundred percent after a final review gate. Each step should require a positive quality signal, not just the absence of a negative one. The distinction matters: a system that is not detectably worse than the prior version is not the same as a demonstrably better system. Graduation criteria that require a positive signal reduce the risk that an undifferentiated change accumulates hidden technical debt across multiple releases.

12.8 Shadow Deployments for Quality Comparison

Shadow deployment is the single most-leveraged quality pattern available to a team releasing LLM changes. In shadow mode, every production request is routed to both the current and candidate models. Only the current model's response is returned to the user; the candidate model's response is captured, scored, and stored for analysis. The user experience is entirely unchanged, and the team accumulates a large, production-representative comparison corpus before any user sees the new model.

The infrastructure requirements for shadow deployment are modest. The routing layer fans out each request to both model endpoints, collects both responses, returns the current model's response immediately, and asynchronously writes both responses to a comparison store. The asynchronous path does not affect production latency. The comparison store feeds an analysis pipeline that scores both responses using the eval rubric, aggregates the scores by traffic category, and produces a daily summary report. Shadow infrastructure typically requires one to two days of engineering effort to build, and the return on that investment is captured on every subsequent release.

Shadow Deployment Architecture

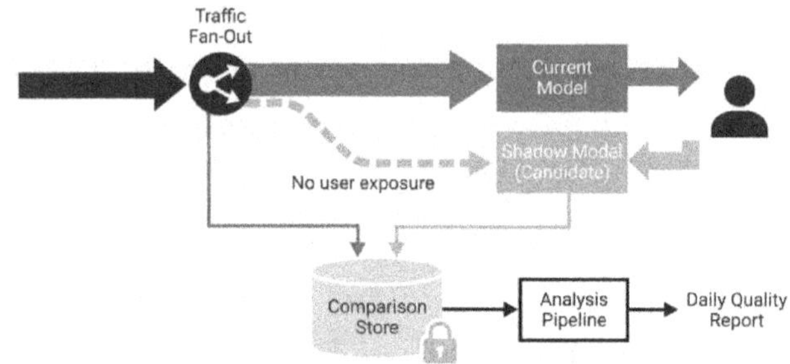

Shadow scoring at scale requires automated quality measurement rather than human review. The same rubric-based scoring approach used in the eval suite applies here: each captured pair of responses is scored on the relevant rubric dimensions, and the scores are stored alongside metadata (traffic category, session length, user cohort, timestamp) that enables segmented analysis. The segmentation is where most of the value lives. An aggregate score improvement that masks a regression in a high-value user segment is exactly the failure mode that shadow deployment is designed to surface before it reaches production.

A shadow campaign runs for a defined period before a release decision is made. The appropriate duration depends on the traffic volume needed to achieve statistical significance across all segments, the degree of concern about specific edge cases, and the urgency of the release. For high-traffic systems, 24 to 48 hours of shadow data typically provides sufficient volume for a confident release decision. For lower-traffic systems, the campaign may need to run for several days or cover specific traffic scenarios by replaying the historical corpus in addition to capturing live traffic.

Side-by-side review of shadow output pairs is the qualitative complement to automated scoring. After the automated analysis identifies categories of divergence between the current and candidate models, a human reviewer samples pairs from those categories and reads both responses directly. This review often surfaces differences that automated rubrics miss: changes in the model's level of explanation, shifts in hedging behavior, alterations in the handling of ambiguous questions, or changes in the style of refusals. The side-by-side review is not a replacement for automated scoring; it is the qualitative signal that decides close cases where automated scores are within the noise band.

12.9 Reproducible Training and Adaptation Pipelines

A fine-tuned model or a LoRA adapter is an artifact. Like any artifact in a well-run engineering organization, it should be reproducible from a documented set of inputs: the base model version, the training dataset version, the hyperparameter configuration, the training infrastructure specification, and the evaluation results that certified it for release. When any of these inputs is undocumented or unavailable, the artifact cannot be reproduced, audited, or compared meaningfully to a future version.

Training pipelines as code is the foundational pattern. The training procedure should be expressed as a reproducible pipeline definition that specifies every step from dataset loading through evaluation: data preprocessing transforms, tokenization settings, batch construction logic, loss function configuration, optimizer settings, learning rate schedule, checkpoint cadence, and post-training evaluation. Pipeline definitions live in version control alongside the rest of the codebase, and every training run is launched from a specific, tagged version of that definition. This makes every training run an instance of a versioned procedure rather than a manual process that relies on the engineer who ran it.

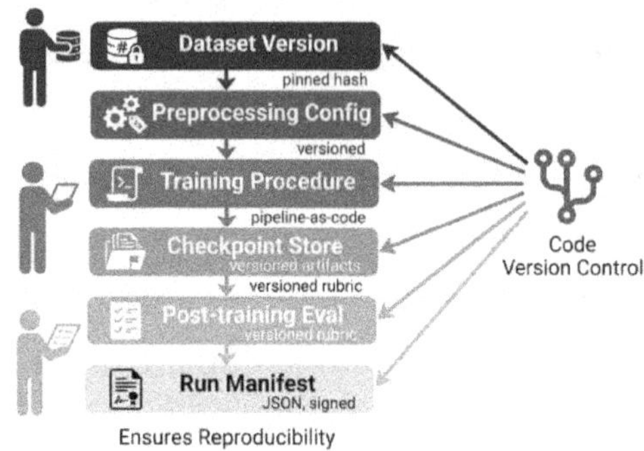

Dataset versioning requires more care than code versioning because training datasets are large, often proprietary, and frequently modified between training runs. The minimum requirement is a content hash pinned to each training run so that the exact dataset state used for training can be verified later. Beyond the hash, the dataset version record should capture the provenance of each data source, the filters applied during construction, the deduplication method, and the rationale for the train/validation split. Teams that collect feedback from production to augment training datasets must version the feedback-incorporation procedure with the same rigor as the original dataset construction, because production-feedback data introduces temporal dependencies that affect reproducibility.

The run manifest is the document that ties everything together. It records the base model identifier (including the provider's version tag or a content hash for self-hosted models), the pipeline definition version, the dataset version, the hardware configuration, the random seed, the final evaluation results, and the artifact storage path for the trained checkpoint or adapter. The run manifest is signed and stored alongside the artifact; it is the artifact's birth certificate. A team that can hand an auditor a run manifest for every model in production has a fundamentally different compliance posture than one that cannot.

12.10 Artifact Storage and Model Registry Patterns

An LLM application in production produces more types of artifacts than a classical machine learning system. The artifact inventory includes base model weights (or a reference to the provider's versioned endpoint), fine-tuned model checkpoints, LoRA or other adapter files, prompt templates (which are versioned configuration artifacts, not static files), evaluation suites (the test cases and rubrics used to certify a release), and evaluation results (the scores and analysis produced by running those suites). Each artifact type has a different lifecycle, size profile, and set of consumers. Managing them together in a coherent registry is the structural precondition for artifact lineage. Architects set the durability boundary: which artifacts are immutable for which retention windows, which environments can read them, and how the registry connects to the broader platform's identity and policy planes.

Artifact lineage is the graph that connects each production deployment to the artifacts it depends on and to the evaluation evidence that supports it. A lineage graph for a typical LLM deployment has edges connecting the deployed endpoint to the base model version. This adapter modifies it, the prompt templates that govern its behavior, the evaluation suite that certified it, and the evaluation results from the most recent run of that suite. When a quality regression is detected in production, the lineage graph identifies what changed between the current deployment and its predecessor.

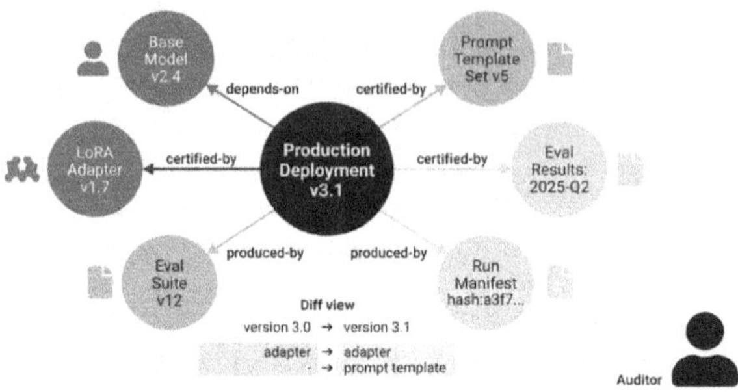

Model registry design for LLM applications typically extends a standard ML model registry with prompt and adapter management capabilities. The registry needs to support semantic versioning for all artifact types, lifecycle states (development, staging, production, deprecated), and access control policies that prevent unreviewed artifacts from being deployed to production. The prompt template registry is often the most neglected component because prompts look like configuration rather than code. Treating prompts as first-class versioned artifacts (with review gates, staging environments, and rollback capability) is one of the highest-leverage structural improvements a team can make to its release safety posture.

Retention and eviction policies for model artifacts require explicit decisions, given the high storage costs. A full fine-tuned checkpoint for a large model can occupy hundreds of gigabytes; a LoRA adapter is much smaller, but the policy for how many adapter versions to retain still needs to be deliberate. The minimum retention requirement is the ability to redeploy any version that was ever in production, typically meaning retaining the two or three most recent production versions, plus any version referenced by an open compliance audit. Intermediate development checkpoints can be evicted aggressively once the run has been promoted or abandoned.

12.11 Continuous Evaluation in Production

Pre-production testing validates a model change before it reaches users. Continuous evaluation validates that the model continues to perform correctly after it reaches users and as the world changes around it. User behavior shifts, new question categories emerge, adversarial users probe the system for weaknesses, and the factual world the model describes evolves. A model that was accurate and safe on release day can drift into inaccuracy or safety violations without any code change, simply because the distribution of inputs has shifted or the domain ground truth has changed.

The continuous eval pipeline samples a percentage of production traffic on an ongoing basis, scores the sampled outputs using the eval rubric, and aggregates the scores into time-series metrics. The sampling rate is a cost-quality trade-off: a higher rate yields more sensitive drift detection, but at a higher evaluation cost. A common approach is to sample at a rate that achieves statistical significance on the most important traffic categories within a twenty-four-hour window, typically one to five percent of total traffic for high-volume systems.

Continuous Evaluation in Production: The Feedback Loop

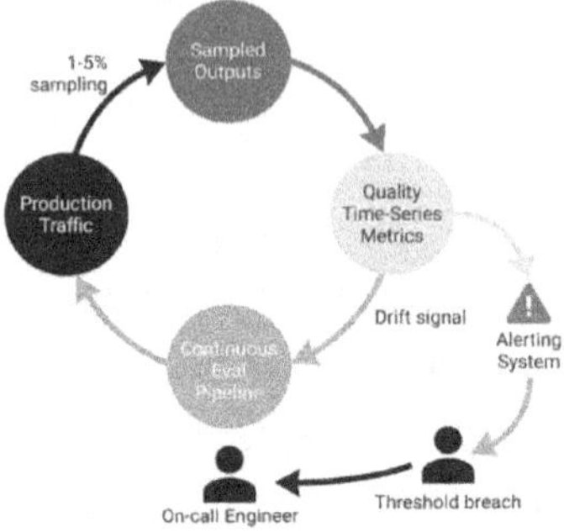

Drift detection in continuous eval has two dimensions: performance drift and distribution drift. Performance drift is a decline in eval scores over time: the model is producing lower-quality outputs than it did at release. Distribution drift is a change in the distribution of input types: the traffic profile has shifted toward categories the model handles less well, or toward entirely new categories not covered by the eval suite. Distribution drift is harder to detect because it does not immediately degrade scores; it degrades coverage. A team that monitors only scores will miss a distribution drift until it has already caused a coverage gap large enough to produce visible quality problems.

Wiring continuous eval to the same alerting paths as operational metrics is the design decision that makes it effective. If latency alerts and eval score alerts route to different on-call rotations with different response expectations, eval regressions will be treated as second-class concerns relative to operational incidents. Configuring eval threshold breaches to produce incident tickets with the same severity classification as latency threshold breaches communicates to the team that quality regressions are incidents, not research artifacts. This is the cultural alignment that separates teams with operational clarity from teams that rely on customers to catch their quality regressions.

Feedback integration connects continuous eval back to the training and evaluation pipeline. When continuous eval identifies a segment of production traffic in which quality is consistently below the threshold, that segment is a candidate for targeted data collection and fine-tuning. When it identifies a category of inputs not covered by the eval suite, that category is a candidate for suite expansion. The continuous eval pipeline is not just a monitoring tool; it is the signal that drives the improvement cycle. Teams that close the loop from continuous eval to training data to eval suite expansion build a compounding quality improvement process rather than a one-time certification ceremony.

12.12 Release Engineering for LLM Applications

Release engineering for LLM applications adds several concerns not present in classical service deployments. The release unit is not just a service binary; it is the combination of a model version, an adapter version, a prompt template set, and a configuration that governs their interaction. Any of these components can change independently, and any change to any component constitutes a release that requires the full testing and deployment discipline. Teams that treat prompt template changes as configuration updates rather than releases consistently underinvest in the testing and rollback infrastructure those changes need.

Feature flags applied to LLM releases allow the team to decouple deployment from activation. A new model version can be deployed to the serving infrastructure before any user sees it, then activated for specific user cohorts under a feature flag. The flag controller enables the team to activate for internal users first, then for a small external cohort, then for progressively larger cohorts, and to deactivate instantly at any stage. Feature flags require a flag management system with audit logging; without an audit trail, it becomes impossible to reconstruct which model version a given user was exposed to at a given time, which is both an operational and a compliance problem.

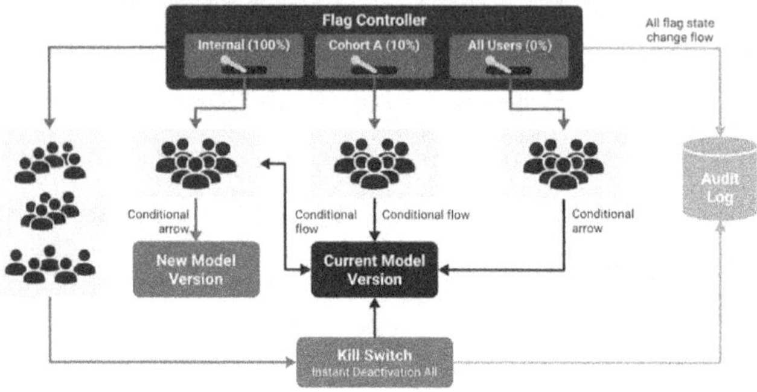

Tenant cohort management extends feature flags to multi-tenant systems where different customers have different sensitivities to model changes. An enterprise customer may require contractual notification before a model upgrade; a self-serve customer may receive updates as part of the standard release cadence. The flag system needs to support both granularities: global rollouts by percentage and targeted activations by tenant identifier. Release tooling that conflates these two dimensions creates operational risk when tenant-specific commitments are not honored.

Kill switches are the emergency variant of feature flags: single-action controls that immediately revert all users to the prior model version without requiring a code deploy or artifact replacement. A kill switch is distinct from a flag because its activation path is hardened for speed and reliability: it should work even if the flag management system is under load, it should not require database writes, and it should take effect within seconds rather than minutes. Building and testing kill switches is a release engineering responsibility, not an incident-response improvisation. An untested kill switch is not a safety mechanism; it is a liability.

Rollback-by-default is the design principle that governs every other release engineering decision in this section. Every release must be rollback-able without engineering intervention during business hours. This means the prior artifact versions must be retained and immediately servable, the routing configuration must revert cleanly when a flag is toggled, and the rollback procedure must be documented, tested in staging, and executable by the on-call engineer without specialized domain knowledge. When a team has to reconstruct a rollback procedure during an incident, the incident duration extends from minutes to hours. When rollback is a documented, rehearsed procedure, the incident resolves before most users notice.

12.13 A CI/CD Checklist for LLM Applications

The patterns in this chapter form a pipeline. The CI/CD checklist is the gate at the end of that pipeline. A team that can check every item on the list before declaring a release production-ready has built the structural discipline that separates confident shipping from hopeful shipping. The checklist is not a compliance artifact; it is a shared agreement about what the team considers a finished release.

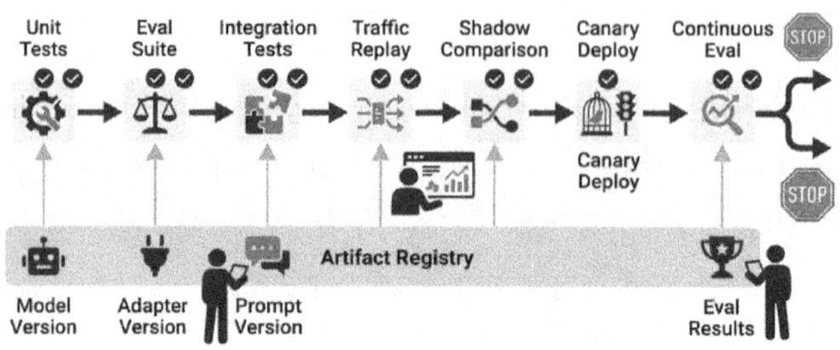

The checklist below synthesizes the required elements of a production-grade LLM release process. Each item represents a gap that has led to a production incident in real organizations operating LLM applications at scale. The items are ordered by the release pipeline phase they apply to, from pre-commit through post-deployment.

The pre-commit layer covers the fast, local checks that run on every code change: prompt snapshot tests pass with no unexpected diffs, orchestration unit tests pass in mock mode, output parsing and schema validation tests pass, and no em-dashes or other formatting regressions appear in rendered prompts. The pull-request layer covers the integration checks that run when code is proposed for merge: contract tests pass between all pipeline components, end-to-end integration tests pass in the staging environment, and any change to a prompt template includes a rendered diff in the review description.

LLM Release Readiness: The Decision Tree

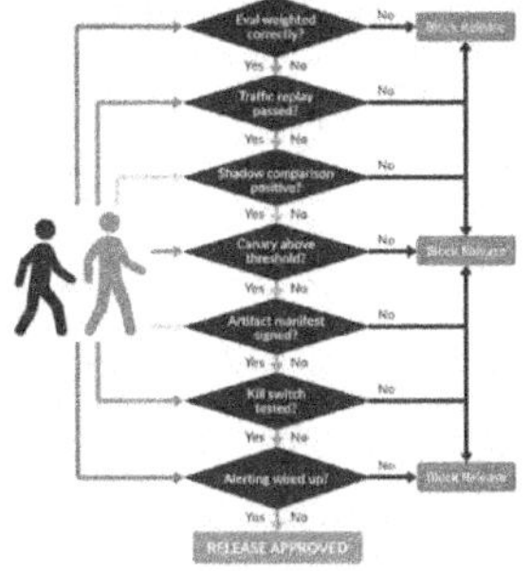

The pre-release layer covers the quality gates that run before traffic is shifted: the eval suite passes with weighted coverage above threshold, the traffic replay comparison shows no significant regressions in high-traffic categories, and the shadow deployment campaign has accumulated sufficient volume for a statistically significant release decision. The deployment layer covers the safety mechanisms that govern traffic shifting: the canary is configured with quality guardrails (not just operational guardrails), the feature flag or canary configuration has been reviewed and documented, and the rollback procedure has been confirmed executable in staging within the current release window.

The post-deployment layer covers the monitoring and closure steps that run after traffic has shifted: continuous eval is running and wired to alerting, the artifact manifest has been signed and stored, the release record links the deployment to the artifact manifest and eval results, and the prior model version has been confirmed retained and servable for rollback. A release that completes all of these steps is an auditable event, not just an operational one, and it creates the foundation for the next release to be incrementally better rather than just incrementally different.

12.14 Technical Checklist

1. Eval suites are versioned and weighted to reflect the shape of production traffic, with weights updated at least monthly.
2. Traffic-replay testing is a required pre-production gate for all high-impact model, adapter, and prompt changes.
3. Canary, blue-green, or shadow deployments are structurally required (not optional) for every model and prompt change.
4. LLM-as-judge evaluators are calibrated against human review on a documented cadence of four to eight weeks.
5. Training and adaptation pipelines are expressed as code in version control and reproducible from a run manifest alone.
6. Run manifests are signed and stored alongside artifacts; every production model version has a corresponding manifest.
7. Artifacts (models, adapters, prompts, eval results) are tracked in a registry with explicit semantic versioning and lifecycle states.
8. Artifact lineage is maintained as a graph connecting each production deployment to its model, adapter, prompt, and eval artifacts.
9. Continuous evaluation is wired to the same alerting infrastructure as operational metrics, with equivalent severity classification.
10. Continuous eval covers both performance drift (score decline) and distribution drift (shift in input categories).
11. Feature flags and tenant cohort controls support both global percentage rollouts and targeted tenant-level activations.
12. Kill switches are implemented, documented, and tested in staging before every production release.
13. Rollback procedures are documented, rehearsed, and executable by the on-call engineer without domain specialist intervention.

14. Mock-mode unit tests cover all branches of the orchestration logic without requiring live-model API calls.
15. Prompt template snapshot tests are in version control and fail when rendered output changes unexpectedly.
16. Schema validation tests cover all structured output contracts between pipeline components.
17. Shadow campaigns run for sufficient duration and volume to achieve statistical significance before a release decision is made.
18. Release records link each deployment event to its artifact manifest, eval results, and prior version for full auditability.

12.15 Team Conversation

1. When did our eval suite last fail to catch a regression that customers caught first? What category of input was involved, and is that category now weighted appropriately?
2. Does our current eval reflect the shape of production traffic, or is it derived primarily from a public benchmark? What would it take to reweight it?
3. Have we ever shipped a model change via shadow deployment? If not, what has been the obstacle, and which release on the roadmap is the right candidate to pilot the pattern?
4. Could we reproduce our most recent fine-tuned adapter from the artifacts in storage today? What is the first step in the run manifest that we cannot reconstruct?
5. Where in our artifact chain is lineage broken? Can we trace the current production deployment back through model, adapter, prompt, and eval to a signed run manifest?
6. Is our LLM-as-judge calibrated against human review, or are we using it as a gate without measuring its agreement rate? What would a calibration run cost in time and reviewer effort?

7. What is the slowest gate in our current CI/CD pipeline, and does its slowness reflect genuine quality validation or accumulated process debt?
8. Have our kill switches been tested end-to-end in staging? Does the on-call engineer know how to activate one without escalating to the ML team?
9. What is the on-call response expectation for a continuous eval threshold breach? Is it documented with the same clarity as a latency SLO breach?
10. Which of our current prompt templates is treated as configuration rather than as a versioned artifact with a release gate? What is the risk profile of that decision?

12.16 Key Takeaway

Testing LLMs is its own engineering discipline. Classical patterns (unit tests, integration tests, CI/CD pipelines) are necessary but not sufficient for systems that produce non-deterministic outputs, where quality is not a binary pass/fail, and where the distribution of inputs shifts continuously in production. The team that adapts those classical patterns with soft assertions, rubric-based eval, traffic weighting, shadow deployments, and continuous eval wired to alerting has built a system that gives them real confidence in every release. The team that does not adapt them has built a system that gives them the appearance of confidence while hiding the regressions their customers will eventually find.

Artifact lineage is the connective tissue of the entire discipline. A model version that cannot be traced to a signed run manifest, a deployment that cannot be linked to its eval results, a prompt template that has no version history: each of these gaps is both an operational liability and a compliance exposure. Building the artifact registry and lineage graph is not glamorous engineering work, but it is the infrastructure that makes audits survivable, incident investigations fast, and quality improvements compounding. A team with a complete lineage graph for every production artifact understands its system in a way that a team without one cannot, regardless of how sophisticated its tooling is.

The CI/CD checklist at the end of this chapter is the pilot-to-production pathway for the patterns described throughout it. A team that treats the checklist as a release gate rather than a reference document has operationalized the discipline. Every item on the checklist represents a failure mode that has produced a real incident; every item that is checked off represents a failure mode that has been closed. Responsible by design means the engineering process itself prevents the failure, not that the team responds well after the failure occurs. That is the difference between a production-grade LLM system and one that happens to be running in production.

13 Human-in-the-Loop and Feedback Systems

13.1 Opening Scenario

The internal knowledge assistant had been running in production for fourteen months when the LLM platform team decided to use the accumulated feedback for the first fine-tuning cycle. The assistant served roughly three hundred staff across two business units, and a customer-success engineer had instrumented a thumbs-up and thumbs-down widget from day one. By the time the team sat down to audit the dataset, they had logged just over forty thousand feedback events. The numbers looked promising on the surface.

The audit took four hours and produced a series of uncomfortable findings. First, 31% of all feedback events came from eight power users who rated generously and quickly, skewing the dataset toward outputs those individuals happened to prefer. Second, the thumbs-down signal covered at least six distinct failure modes: factual errors, formatting complaints, hallucinated policy references, overly long responses, responses that ignored context already in the conversation, and off-topic completions. Every one of these had been flattened into a single binary bit. Third, the labeling rubric that the early team had written in a shared document had been quietly edited on three occasions as the assistant scope expanded, with no version tracking and no re-annotation of older labels from prior rubric versions. The dataset was effectively a mix of answers to three different questions, labeled by a non-representative population, with the negative class carrying no diagnostic information.

The team could not fine-tune it. More labels would not fix structural problems with the feedback architecture. What the team needed was a versioned rubric, a labeling pipeline with active-learning prioritization to direct annotation budget at the cases that actually mattered, a UX layer that elicited structured signal alongside the binary vote, and a documented cadence for turning user signals into model updates. This chapter is the architecture the team built going forward: the canonical reference for human-in-the-loop systems that close the loop from production behavior back to the next model release.

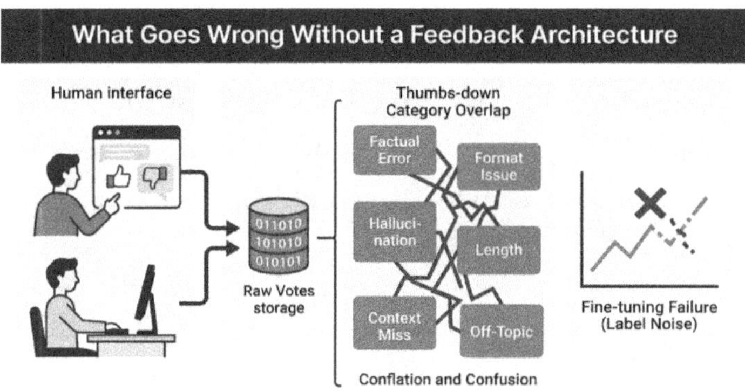

13.2 Why It Matters

Feedback is the only mechanism that connects production behavior to the next model release. Without it, the model does not learn from real usage; it only learns from the dataset used for initial training. That gap widens over time as the deployment context shifts, the user population grows, and the edge cases the model encounters diverge from the training distribution. A system without a functioning feedback loop decays relative to its environment, even when no one changes the code. The workflow-level impact of a missing or broken feedback loop is measured in months of compounding drift before the degradation becomes obvious.

The feedback architecture also determines what a team can know about its model. Implicit signals, such as regeneration events, copy actions, and follow-up clarifications, indicate which outputs users acted on and which they abandoned. Explicit structured ratings tell a reviewer which specific dimension of quality failed. Escalation logs indicate to the safety team which requests categories the model is not ready to handle alone. None of this signal is free. Each stream requires instrumentation, storage, a schema, and a consumer. Teams that treat feedback as an afterthought collect it carelessly and then discover, as the team in the opening scenario did, that the data they spent a year accumulating is not usable. Responsible-by-design feedback collection is a first-class engineering problem, not a product analytics add-on.

The operational stakes are highest where human review sits in the critical path. A poorly designed review workflow burns reviewer time on low-value cases, starves high-value cases of attention, and creates the conditions for reviewer fatigue that degrades labeling quality invisibly. A well-designed workflow routes cases efficiently, calibrates reviewers regularly, and retires human review from workflows where the model has demonstrated sufficient reliability. This chapter covers every layer of that system, from the rubric that defines what reviewers are measuring to the data flywheel that returns their judgments to the model. The pilot-to-production pathway for any LLM feature with meaningful quality requirements runs through a functioning feedback system.

13.3 Labeling Pipelines and Active Learning

A labeling pipeline is the infrastructure that transforms raw model outputs into a structured training signal. At a minimum, it consists of a rubric that defines what annotators should measure, a queue that surfaces items for review, tooling that presents items in a consistent format, and a storage layer that records judgments with enough metadata to reconstruct the labeling context later. At the production scale, it also includes inter-annotator agreement measurement, active-learning prioritization, and a versioning system that ties every label to the rubric version under which it was created.

The rubric is the load-bearing element. It defines the quality dimensions that matter for the specific deployment, the scale points and their meanings, the edge-case guidance that prevents annotators from interpreting ambiguous cases differently, and the examples that anchor each scale point. A rubric without worked examples produces high variance across annotators. A rubric without a version identifier produces a dataset that quietly mixes judgments made under different standards. Both problems are invisible in aggregate statistics and expensive to detect later.

Inter-annotator agreement (IAA) is a measure that tells a team whether its rubric is producing a consistent signal. Cohen's kappa for categorical labels and Krippendorff's alpha for ordinal or continuous ratings are the standard choices. An IAA below 0.6 for a categorical dimension indicates that the dimension is ambiguous in the rubric, the annotators have not been adequately trained, or both. Teams should measure IAA on a random sample of doubly annotated items every week, not just at the start of a labeling campaign. An agreement that is high at launch can degrade over time as annotators develop independent interpretations of edge cases.

Labeling Pipeline Architecture

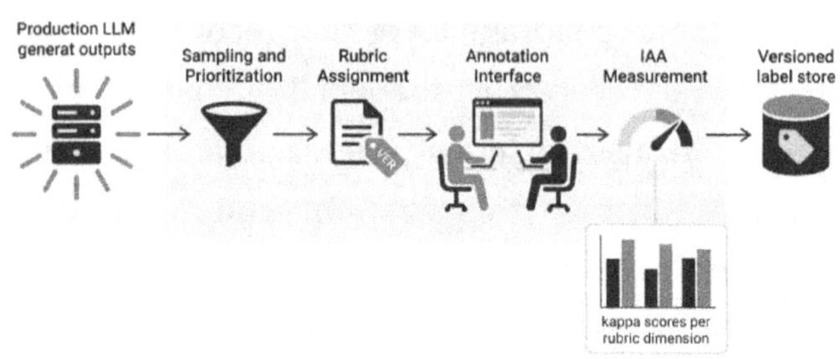

Active learning directly addresses the annotation budget problem. Not all model outputs are equally informative for labeling. Outputs on which the model is highly confident and already correct contribute little when labeled; outputs near the model's decision boundary or in regions of high uncertainty contribute much more. An active-learning prioritization layer scores the queue by informativeness before surfacing items to annotators, directing budget toward the cases where labeling effort produces the greatest reduction in model uncertainty.

The three most practical active-learning strategies for LLM labeling pipelines are uncertainty sampling, which surfaces outputs with the lowest model confidence scores; diversity sampling, which ensures coverage of semantically distinct request types; and error-signal sampling, which prioritizes items that triggered negative implicit signals, such as regeneration or abandonment. In practice, a weighted mix of all three outperforms any single strategy because uncertainty alone tends to surface idiosyncratic edge cases while diversity alone ignores the model's actual failure modes.

13.4 RLHF, DPO, and Preference-Tuning Patterns

Reinforcement learning from human feedback (RLHF) is the preference-tuning pattern that shaped the first generation of instruction-following LLMs. The workflow has three stages: collect pairwise preference judgments from human annotators who compare two model outputs and indicate which is better; train a reward model on those judgments so it can score any output; and use that reward model as the signal for a reinforcement-learning fine-tune of the base LLM. The appeal is that human preference is a richer signal than a scalar loss function. The cost is operational: running RLHF at production scale requires a stable reward model, a reinforcement learning training loop with careful KL-divergence constraints to prevent the policy from drifting too far from the base, and a human annotation pipeline capable of producing pairwise judgments at the volume and quality required by the reward model.

RLHF Three-Stage Workflow

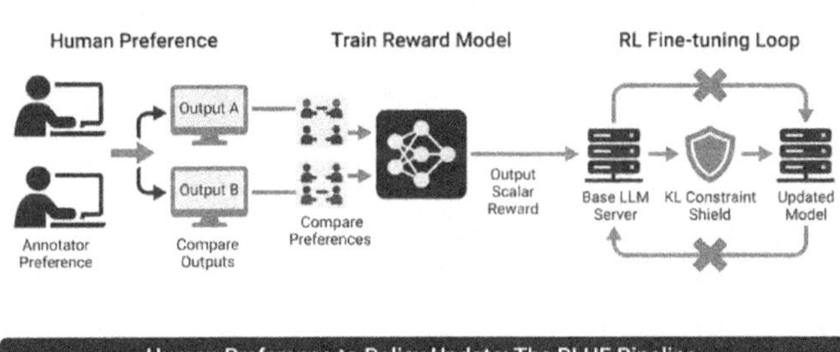

Human Preference to Policy Update: The RLHF Pipeline

Direct preference optimization (DPO) removes the need for an explicit reward model. Instead of training a reward model and using it to drive RL, DPO directly optimizes the language model's parameters against the pairwise preference dataset using a closed-form loss derived from the Bradley-Terry preference model. The practical effect is that teams can run preference tuning without maintaining a separate reward model, without a reinforcement learning loop, and without the instability that RL training can introduce. DPO is now the preferred pattern for most production preference-tuning workflows because its training pipeline is simpler, its behavior is more predictable, and its compute requirements are lower.

The operational distinction between an RLHF or DPO experiment and a production preference-tuning pipeline is the same distinction that separates any ML experiment from a production system. Production pipelines have versioned preference datasets with IAA thresholds as a data quality gate, automated retraining triggers when new preference data accumulates past a configured batch size, evaluation benchmarks that run after every retrain and must pass before the update is promoted, and rollback procedures that can revert a policy update within a defined time window. Experiments have none of these. Teams that run preference tuning ad hoc, accumulating labels whenever convenient and retraining on manual schedules, will find that the quality improvements they observe in controlled conditions do not reliably carry over to production.

13.5 Human Review Workflows and Escalation Paths

Human review in an LLM system serves two distinct functions that are easy to conflate. The first is quality control: a human reviews an output before it reaches the user or is stored, catches errors the model made, and either corrects the output or flags it for follow-up. The second is data collection: a human labels an output to generate a training signal that improves the next model version. Both functions require human attention, but they have different latency requirements, different reviewer skill requirements, and different downstream consumers. Conflating them produces a workflow that does neither well.

Tiered escalation separates cases by the urgency and expertise they require. A three-tier structure covers most production deployments. Tier one is automated triage: a classifier or rule set identifies outputs that require review and routes them to a queue without human involvement. Tier two is first-line review: a reviewer with domain context but not subject-matter expertise reviews the output against the rubric, approves it, corrects it, or escalates it. Tier three is expert review: a subject-matter expert handles cases that first-line review could not resolve, outputs in high-stakes categories such as legal or medical content, and outputs that pattern-match to known safety categories.

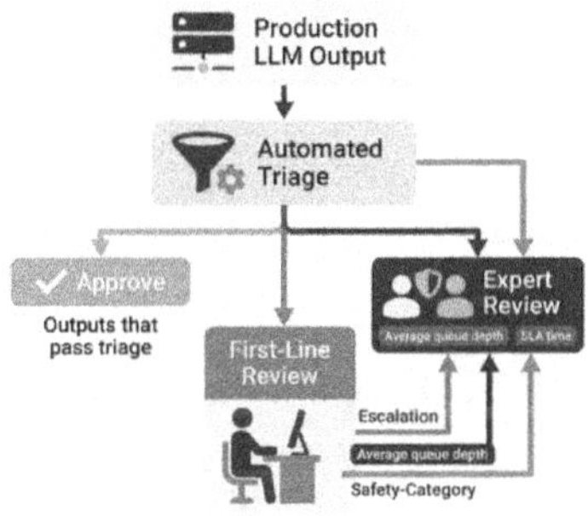

Tiered Escalation Matches Case Complexity to Reviewer Skill.

Reviewer burnout is a structural risk that degrades labeling quality in an invisible way. Reviewers who process too many cases per session tend to make faster, less careful judgments. Reviewers who see too many similar cases develop anchor bias, applying the first judgment pattern that worked to subsequent cases regardless of differences. Reviewers who receive no feedback on their calibration drift from the rubric silently. The operational patterns that prevent burnout include daily case-volume caps, session-length limits, mandatory calibration exercises inserted as gold questions into the review queue, and regular individual feedback sessions in which reviewers see how their judgments compare to calibrated standards.

Ownership matters more than tooling. A review queue with no named owner becomes a backlog—escalation paths with no named resolver at tier three become sinks where hard cases disappear. The operational clarity requirement for human review is simple. Every queue has an owner, every escalation path has a named resolver at each tier, and both are reflected in the team's runbook with documented SLAs. This is not organizational overhead. It is the prerequisite for the feedback system to function when the model produces a novel failure mode at two in the morning.

13.6 UX Patterns for Feedback Collection

The UX of a feedback collection surface determines how much signal the system captures, from whom, and under what conditions. A binary thumbs-up widget at the bottom of a chat response is the most common implementation and also one of the least informative. It captures a global sentiment but discards the dimension of quality that drove it. An engineer looking at a dataset of thumbs-down votes cannot tell whether users were frustrated by accuracy, format, response length, or tone without additional instrumentation.

Structured follow-up prompts improve signal quality with modest UX cost. The pattern works as follows: after a user provides a negative rating, the interface presents a short, structured question that takes one click to answer. For a knowledge-assistant deployment, the options might be: this was factually wrong, this missed the point of my question, this was formatted poorly, or something else. The click adds one dimension of structured signal to the binary vote without requiring the user to type anything. Structured follow-up prompts require rubric alignment between the UX design and the labeling system; the categories presented to users should map directly to the rubric dimensions reviewers are annotating against.

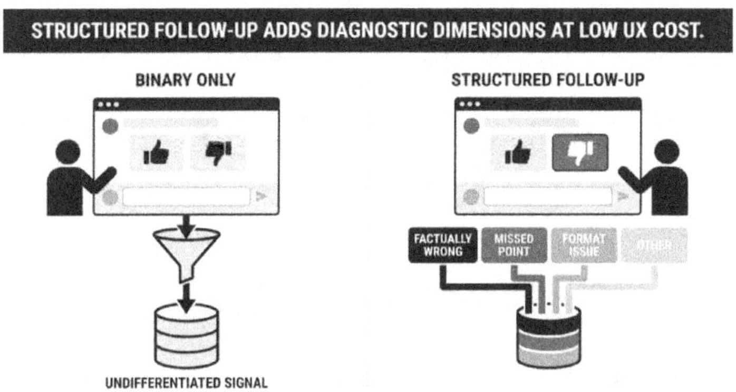

Feedback yield is the fraction of outputs that receive any feedback signal. Binary widgets on chat interfaces typically yield below 5%, meaning that more than 95% of outputs produce no signal. Techniques for increasing yield include contextual prompting, which surfaces the feedback widget more prominently on outputs where the user paused, re-read, or scrolled; session-end surveys, which ask about quality at the end of a session rather than after each output; and periodic targeted requests, which ask users to rate a sample of their recent outputs on a documented cadence. Each approach has tradeoffs: contextual prompting introduces selection bias toward outputs that triggered unusual user behavior, session-end surveys suffer from recency effects, and targeted requests can feel intrusive if overused.

Respecting non-participating users is both an ethical requirement and a system design constraint. A feedback collection surface that nags users who have not engaged with it produces low-quality coerced responses and erodes trust in the product. The correct design treats feedback as opt-in at the session level: users who engage with the widget receive more of it; users who never engage receive a small number of unobtrusive prompts and are then left alone. This produces a smaller but higher-quality signal from users who have chosen to participate, which is more valuable than a larger but lower-quality signal from users who felt pressured.

13.7 Implicit Feedback Signals: Regenerate, Copy, Edit, Follow-Up

Implicit feedback signals are the behavioral traces that users leave without being asked. They are typically more abundant than explicit ratings by an order of magnitude or more, and they are unaffected by the response biases that distort explicit ratings. A user who copies a model output to a document, continues working from it, and never opens a new query has implicitly signaled that the output was useful. A user who immediately hits the regenerate button has implicitly signaled that it was not. A user who types a follow-up clarification has implicitly signaled that the response was incomplete or ambiguous. None of these signals requires the user to do anything they would not have done anyway.

The four primary implicit signals for text-generation systems are regeneration rate (the fraction of outputs that trigger a regeneration request), copy rate (the fraction of outputs from which the user copies text), edit rate (the fraction of outputs that the user modifies before using), and follow-up clarification rate (the fraction of outputs that are followed immediately by a clarifying question). Each signal measures a different dimension of output quality. Regeneration rate captures outright rejection. Copy rate captures utility. Edit rate captures partial utility with quality gaps. The follow-up clarification rate captures completeness and clarity of communication.

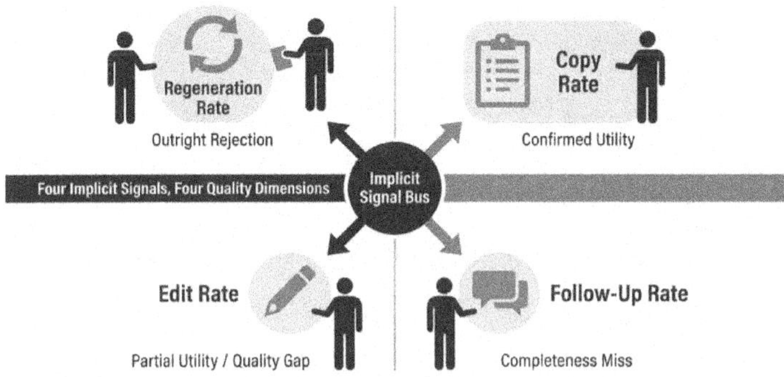

Instrumentation for implicit signals requires agreement between the product team and the LLM platform team on what events to log and how to join them to the output that triggered them. The minimum required schema includes a session identifier, an output identifier, a timestamp, an event type (regenerate, copy, edit, follow-up), and, for edit events, a representation of the edit distance between the model output and the user's modified version. Edit distance is a proxy for the magnitude of the quality gap: a user who changes three words has a different experience from a user who rewrites the entire response. Both are edits; only the distance captures the difference.

Implicit signals should be treated as the primary source of feedback for continuous monitoring and as a supplement to explicit ratings for training data generation. They cannot fully replace explicit ratings because they do not carry rubric-aligned diagnostic information: knowing that a user regenerated an output does not tell the model why the output was rejected. The most effective feedback architectures use implicit signals for high-frequency, low-latency quality monitoring and explicit signals for the lower-frequency, higher-quality labeling work that produces training data. The two streams complement each other and should be stored in schemas that make them joinable by output identifier.

13.8 Feedback Calibration and Rater Quality Control

Rater calibration is the process of measuring how well each annotator's judgments align with a calibrated standard and correcting for drift. Without calibration, annotators diverge silently over time. A rater who began the project aligned with the rubric may, after several months of independent annotation, have developed interpretive shortcuts that systematically differ from what the rubric intends. That drift is invisible in individual label records and only becomes detectable through periodic calibration exercises.

The standard calibration mechanism is gold question seeding: a set of outputs with known correct labels, agreed upon by subject-matter experts and rubric designers, is inserted into the review queue at a regular rate of three to five percent of items. Annotators do not know which items are gold questions. After a calibration period, each annotator's accuracy on gold questions is computed and compared to the calibration threshold. Annotators who fall below the threshold receive targeted retraining; their recent labels on non-gold items may be flagged for re-review or down-weighted in the training dataset.

Rater Calibration via Gold Question Seeding

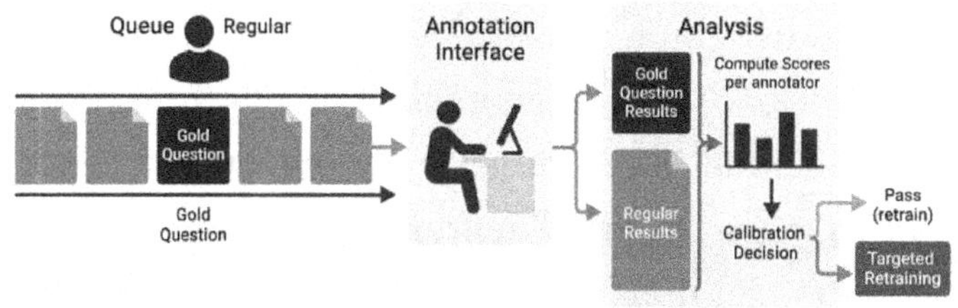

Gold Question Seeding Surfaces Rater Drift Before It Corrupts Training Data.

Rater agreement metrics serve a different purpose from calibration. While calibration measures alignment to a known standard, agreement measures consistency among annotators working on the same items without a known answer. High agreement on ambiguous items indicates that the rubric is clear and that annotators have internalized it consistently. Low agreement on specific item types indicates rubric ambiguity that needs to be resolved through clarifying guidance and additional worked examples, not more annotation.

Rater quality control extends to the detection of gaming behaviors. Annotators who are paid per item have an incentive to rush; annotators who face no accountability for disagreement have an incentive to guess. Quality control mechanisms include minimum time-on-item checks (flagging annotations completed below a duration threshold), consistency checks (presenting the same item twice in a session and measuring self-agreement), and anomaly detection in individual annotator agreement distributions. A rater whose agreement distribution shifts sharply downward over two weeks is a signal worth investigating before that rater's labels enter the training pipeline.

13.9 Closing the Loop: From User Signal to Model Update

The closing-the-loop pattern connects user feedback, through the labeling pipeline and training infrastructure, to a model update that ships to production. Without an explicit architecture for this path, feedback accumulates in storage but never reaches the model. Teams that have collected months of feedback without triggering a single model update have a data flywheel that turns once and stops. The architecture that keeps it turning requires four elements: a feedback accumulation threshold that triggers the next training batch, an automated data preparation pipeline that transforms raw feedback into training-ready format, an evaluation suite that validates the update before promotion, and a deployment gate that prevents a regressing update from reaching production.

The cadence for loop closure should be explicit and documented. A quarterly retraining cycle is common for deployments with moderate feedback volume and stable requirements. A monthly cycle is appropriate for high-volume deployments where quality drift is a significant operational concern. A continuous or near-continuous fine-tuning cycle, sometimes called online learning or shadow fine-tuning, is only appropriate for deployments where the team has invested heavily in automated evaluation and can detect regressions within hours rather than days. The right cadence is the fastest the team can reliably execute, not the fastest technically possible.

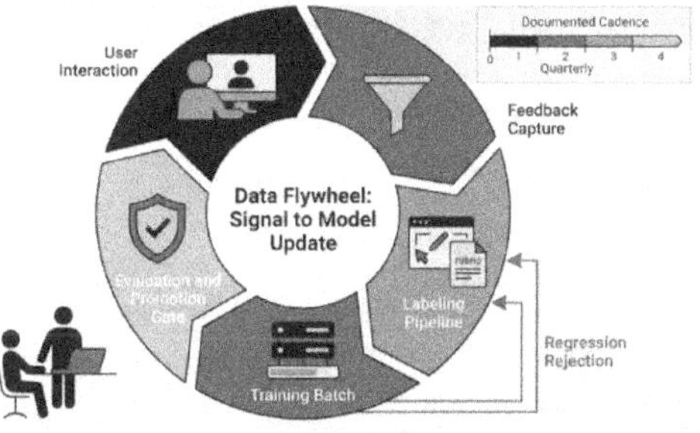

The Feedback Flywheel Requires an Explicit Cadence to Keep Turning.

Bias amplification is the primary risk in a closed feedback loop. If the model's current outputs over-represent certain topics, tones, or response patterns, and user feedback reflects satisfaction with those outputs, retraining on that feedback reinforces the existing biases. The mitigation requires diversity sampling in the feedback collection layer, ensuring that the training batch is representative of the full distribution of request types the deployment handles, not just the request types that generated the most feedback. Stratified sampling by request category, with minimum representation thresholds for each category, is the standard approach.

The evaluation suite that gates loop closure should include at minimum: a set of static benchmark prompts with human-validated expected outputs, a regression test on outputs from the previous model version that were rated positively, and a safety evaluation on a curated set of adversarial inputs. The training batch must improve on the benchmark, must not degrade the positively rated historical outputs, and must pass the safety evaluation before the model is promoted. Evaluation suites that consist only of aggregate metrics can pass while masking regressions on specific request categories that aggregate scores conceal.

13.10 Privacy-Preserving Feedback Collection

Feedback collection in an LLM deployment is a data collection activity subject to the same privacy requirements as any other user data collection activity. This means it requires a lawful basis under applicable data protection law, a disclosed purpose that is consistent with how the data is actually used, a retention policy that is enforced rather than aspirational, and technical measures that prevent user data from leaking into training datasets in ways that violate user expectations or regulatory requirements.

Consent architecture for feedback collection has two layers. The first is ambient disclosure: the deployment's terms of use and privacy notice disclose that outputs may be used to improve the model, giving users the context they need to make informed choices about how they interact with the system. The second is active consent for direct data contribution: if the team wants to use specific user conversations as training data, that use requires separate explicit consent in jurisdictions where data protection law demands it. The two layers serve different purposes and should not be conflated. A privacy notice that buries training data use in a general improvements clause may satisfy the letter of disclosure requirements but fail to meet the spirit of informed consent.

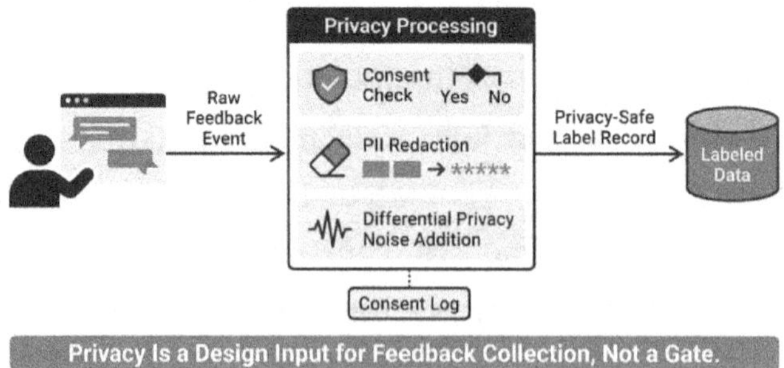

PII redaction is the baseline technical measure for any feedback pipeline that stores conversation content. A redaction layer should strip recognized PII patterns, including names, email addresses, phone numbers, and identification numbers, before the content is stored or transmitted to any downstream system. The redaction layer should be applied at ingestion time, not retroactively, because retroactive redaction requires locating all copies of the unredacted content across all storage systems. This task is operationally difficult and legally risky.

Differential privacy offers a stronger guarantee for aggregate feedback statistics shared across organizational boundaries or published as benchmarks. The technique adds calibrated noise to statistics derived from the feedback dataset such that no individual record can be inferred from the published aggregate. The practical trade-off is accuracy: differential privacy degrades statistical signal quality, with the magnitude of the degradation determined by the privacy budget parameter. For most production labeling pipelines, PII redaction, along with access controls, is sufficient. Differential privacy is appropriate when the aggregate statistics will be shared outside the team's direct control or when the deployment context involves sensitive population groups.

13.11 Cost and Latency Trade-Offs of Human Review

Human review is the most expensive component of a feedback system on a per-item basis. A skilled reviewer processing outputs for a domain-specific assistant may cost an order of magnitude more per item than automated evaluation. They may introduce latency of hours to days, depending on the review queue depth and the reviewer's availability. The engineering problem is not to eliminate human review but to ensure that it is applied where it generates value that automated alternatives cannot and that the deployment's operational requirements bound its cost and latency.

Sampling is the primary cost-control mechanism. Rather than reviewing every output, the team reviews a statistically meaningful sample. For quality monitoring purposes, a one percent random sample often provides sufficient statistical power to detect meaningful shifts in output quality. For training data generation, the sample is replaced by an active-learning prioritized queue that focuses reviewer attention on high-informativeness items. The two use cases have different sampling strategies, reviewer skill requirements, and downstream consumers, and should be managed as separate workflows even when they share reviewer infrastructure.

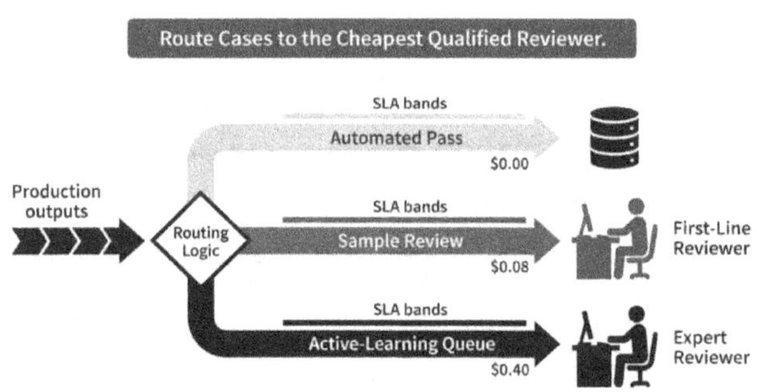

Asynchronous review decouples human-review latency from user-facing response latency in applications where synchronous review is not required by safety or compliance constraints. An asynchronous review workflow logs the output at serving time, routes it to the review queue, and applies the reviewer's judgment retroactively either to correct stored outputs or to generate a training signal. The model's users see the output immediately; the reviewer's judgment arrives later and improves future outputs. Synchronous review, in which the model's output is held pending human approval before delivery to the user, is appropriate only when the cost of delivering an incorrect output is high enough to justify the latency and throughput reduction it imposes.

The routing logic that assigns cases to reviewers should match case complexity to reviewers' skills and costs. A rule-based or classifier-based router can identify cases that require expert review based on topic category, confidence score, or safety classifier output, and send only those cases to the expensive expert queue. First-line reviewers handle the larger volume of standard cases at a lower per-item cost. The cost difference between expert review and first-line review can be an order of magnitude; routing even 5% of cases away from expert review through accurate triage has a significant budget impact at production volume.

13.12 Retiring Human Review: When the Model Is Ready

Human review is a phase, not a permanent state, for most LLM workflows. The goal of a well-functioning feedback loop is to improve model quality to the point where mandatory human review is no longer required for a given workflow, replaced by sampling-based audits that verify the model is maintaining the quality level at which review was retired. Getting to that point requires explicit criteria, not intuition. Teams that retire human review because it feels like the model is good enough, rather than because it has passed documented quality thresholds on a statistically valid sample, tend to discover the quality regression they missed only when a user complaint surfaces.

The explicit criteria for retiring human review from a workflow should include a minimum number of consecutive evaluation periods in which the model's quality score on the review sample has met or exceeded the quality threshold; a minimum sample size for the review periods used to establish that track record; and a maximum acceptable error rate on the rubric dimensions that matter most for the workflow's mission. For a factual-accuracy dimension on a knowledge-assistant deployment, the threshold might be set at a two percent error rate on a weekly one percent sample over twelve consecutive weeks. These numbers are workflow-specific; the principle is that they are written before the review is retired.

Explicit Criteria Replace Intuition for Review Retirement Decisions.

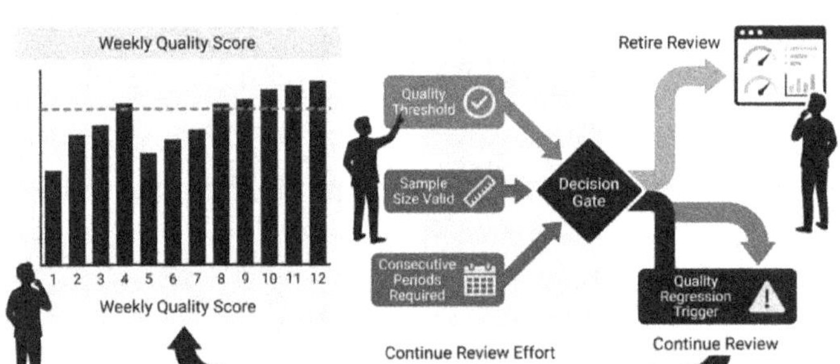

Sampling-based audits replace mandatory review after retirement. The audit reviews a statistically defined sample of outputs on a documented cadence, measures quality against the same rubric used during the review phase, and compares results to the retirement threshold. If the model's quality on the audit sample falls below the threshold for two consecutive audit periods, mandatory review is automatically re-engaged. The re-engagement trigger is part of the retirement architecture, not an afterthought. A team that retires review without a re-engagement trigger has removed its primary quality safeguard without providing a replacement.

Confidence scores can serve as a dynamic review threshold when the model produces calibrated confidence signals. Rather than reviewing a fixed percentage of outputs, the team reviews any output below a confidence threshold, which may be a higher or lower percentage of total volume depending on the model's current confidence distribution. As the model improves and confidence scores shift upward, the review rate naturally decreases. As new request types enter the deployment and push confidence scores down on unfamiliar inputs, the review rate increases automatically. Confidence-based routing requires that the model's confidence scores be well-calibrated; overconfident models with systematically high confidence scores will pass through without review even when their outputs are incorrect.

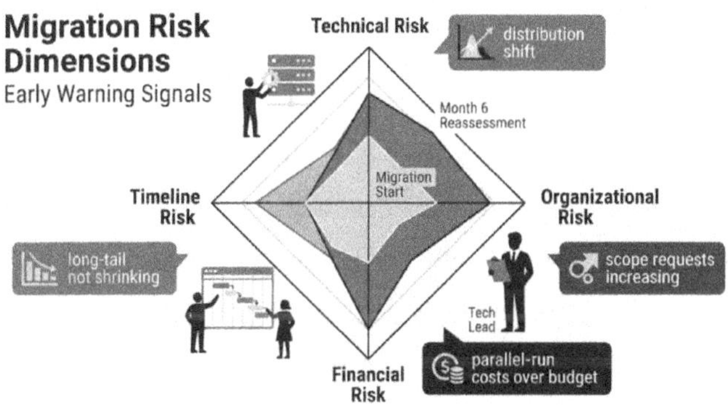

The retirement decision framework applies at the workflow level, not the model level. A single deployment may handle dozens of distinct workflow types, each with its own quality profile and its own review requirements. A billing-inquiry workflow may reach retirement criteria within six months. A contract-drafting workflow in the same deployment may require mandatory review indefinitely because the cost of an error exceeds any tolerable error rate. Managing review requirements at the workflow level, rather than the model level, is the operational clarity pattern that prevents teams from either over-reviewing low-risk workflows or under-reviewing high-risk ones.

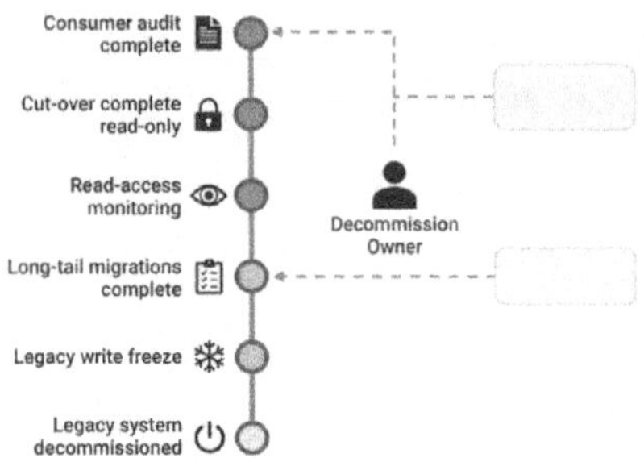

13.13 Technical Checklist

Use the following checklist as a release gate before declaring a system production-ready on human-in-the-loop and feedback requirements.

1. Labeling rubrics are versioned, and each label record carries the rubric version under which it was created.
2. Inter-annotator agreement is measured on a random sample of doubly annotated items each week, with a minimum kappa threshold defined for each rubric dimension.
3. Active-learning prioritization is in place for labeling queues with a finite annotation budget, using a weighted mix of uncertainty, diversity, and error-signal sampling.
4. Preference-tuning pipelines (RLHF or DPO) include automated data quality gates, retraining triggers, evaluation suites, and rollback procedures, all documented in the runbook.
5. Tiered review and escalation paths are documented with named owners at each tier and SLAs for queue depth and response time.

6. Reviewer burnout controls are in place: daily case volume caps, session length limits, and gold question seeding at three to five percent of queue volume.
7. Implicit feedback signals (regenerate, copy, edit, follow-up) are instrumented and stored with schemas joinable to explicit ratings by output identifier.
8. Feedback UX respects non-participating users: opt-in at the session level, with unobtrusive prompting and no coercive patterns.
9. Rater calibration is run on a documented cadence with gold question accuracy thresholds, and annotators below the threshold receive targeted retraining before their recent labels are used in training.
10. A closing-the-loop architecture moves user signals into model updates on a documented cadence with an explicit feedback accumulation threshold, automated data preparation, and evaluation gates.
11. Feedback collection is privacy-compliant: PII is redacted at ingestion time, a consent architecture covers both ambient disclosure and active data contribution, and retention policies are enforced.
12. Explicit quality thresholds, minimum sample sizes, and consecutive-period requirements are documented for retiring mandatory human review from each workflow.
13. Sampling-based audits with an automatic re-engagement trigger replace mandatory review after retirement.
14. Confidence-threshold routing is used only when the model's confidence scores have been validated as well-calibrated.
15. Review cost and routing decisions are documented: which case types go to automated pass, first-line review, or expert review, with cost-per-item and SLA annotations.

13.14 Team Conversation

Use the following questions in team meetings and architecture reviews to pressure-test alignment on feedback system design and operational readiness.

1. If we needed to run a fine-tuning cycle starting today, could we reproduce a clean preference dataset from our current feedback store? What would break, and who would be the first to know?
2. Where does implicit feedback live in our current stack: regeneration events, copy actions, edit distances, follow-up clarification rates? Is any of it being consumed, or is it logged and forgotten?
3. Is our reviewer rubric versioned? If someone changed the rubric six months ago, do we know which labels were collected under the new version and which under the old, or have we mixed them silently?
4. When did we last calibrate our raters? What did the calibration show? Did we act on the findings? What is our gold question seeding rate right now?
5. Which workflows on our current roadmap could graduate off mandatory human review next quarter based on documented quality thresholds? Which will require mandatory review indefinitely, and are those requirements written down somewhere?
6. Is our end-to-end feedback collection privacy-compliant? Does our PII redaction run at ingestion time, or are we relying on retroactive cleaning? Do we have active consent for the use of training data in the jurisdictions where it is required?
7. Is our data flywheel actually turning? When did the last model update incorporate feedback from production? What was the batch size, and what did the evaluation suite show?

8. What is our re-engagement trigger if we retire human review from a workflow and quality regresses? Does it exist in the runbook, or would it require a manual decision under pressure?
9. Do we know the cost per reviewed item for each tier of our review workflow? If our review volume doubles next quarter, does the budget scale with it, and has anyone modeled that?
10. Are we collecting feedback in a way that our users and our compliance team would both be comfortable with if they read the full implementation, not just the privacy notice?

13.15 Key Takeaway

Feedback is the engine of LLM improvement, and engines require maintenance. A feedback architecture that collects binary votes from a non-representative population under an unversioned rubric does not become useful simply by accumulating volume. The signal quality of a feedback dataset is determined by the rubric that defines what annotators are measuring, the calibration system that keeps them measuring it consistently, the active-learning layer that directs annotation budget toward high-informativeness cases, and the data preparation pipeline that transforms raw judgments into training-ready format. Getting those elements right before the first label is collected is far less expensive than correcting them after a year of accumulation, which is the lesson the team in the opening scenario learned the hard way.

Implicit signals are the larger and more reliable source of feedback for continuous quality monitoring. Regeneration events, copy actions, edit distances, and follow-up clarification rates tell the team what users did with model outputs without requiring any additional user action. These signals are unaffected by the response biases that distort explicit ratings, they are available at much higher volume, and they are instrumentable on day one of a deployment without any UX investment beyond event logging. The teams that operate the most effective feedback systems treat implicit signals as the primary monitoring source and explicit ratings as the higher-quality, lower-volume signal reserved for generating training data.

Human review is a phase, not a permanent state, for most workflows in a well-operated LLM deployment. The goal of the feedback loop is to improve model quality to the point where mandatory human review is retired, replaced by sampling-based audits that verify the model is maintaining the quality level at which review was released. Reaching that point requires explicit retirement criteria written down before review is retired, not intuition about whether the model feels good enough. Privacy is a design input throughout this architecture, not a late-stage approval gate. The pilot-to-production pathway for any LLM feature with meaningful quality requirements runs through a feedback system designed from the first day of the pilot, not assembled retroactively when the fine-tuning conversation begins.

14 RAG, Agents, and Tool-Using LLM Systems in Production

14.1 Opening Scenario

The platform team had spent six weeks building what looked, in every demo, like the future. The workflow was called Travel Pilot: an agentic system that accepted a natural-language request from an employee, consulted the internal calendar system to find available windows, queried the corporate travel portal for options, cross-referenced the expense policy document store via retrieval-augmented search, and booked the best match by calling the booking API. The demo ran flawlessly seventeen times. The team gave a 20-minute presentation to engineering leadership, received approval to launch to the first 200 employees, and shipped on Thursday afternoon.

By Friday morning, three employees had duplicate flight bookings. One had five. The incident began when an employee submitted an ambiguous request specifying two cities without a clear direction. The agent entered a loop: it called the booking API, received a response that contained an ambiguity flag the agent had not been programmed to handle, re-queried the policy document store for guidance, received a chunk that was semantically adjacent but not directly applicable, reformulated the booking request, and called the API again. The loop ran 43 times before a monitoring alert in the expense system fired because the total spend code for one employee exceeded the monthly travel budget. By then, three employees had duplicate tickets, and the only record of what the agent had actually done was a generic success log from the booking API.

The postmortem lasted four hours. What the team discovered was not a model problem. The language model had performed roughly as expected on each step. The problem was architectural: the agent had no step budget, no cost ceiling, and no termination condition other than the model deciding the task was complete. The tool integrations were not idempotent: calling the booking API twice for the same request created two bookings rather than returning the first booking's confirmation. The retrieval layer had no mechanism for the agent to distinguish between a document that answered its question and a document that was merely topically adjacent. And the observability surface showed nothing about the agent's internal state during a session: no step trace, no tool call log, no record of which retrieval results had influenced which decisions. The fix was not a smarter model. It was the production architecture that the team should have built first.

Four Reference Architecture Landscape

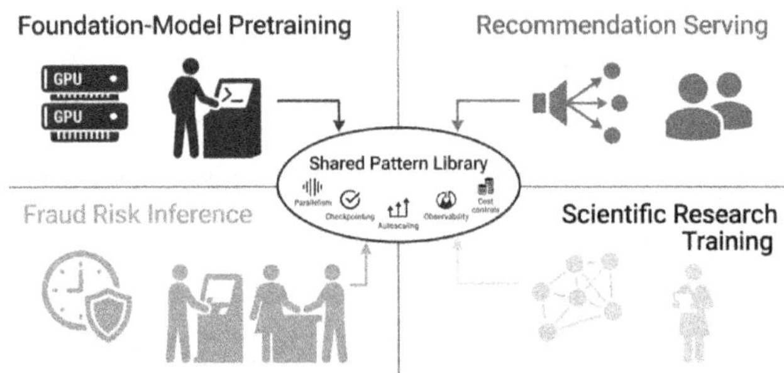

14.2 Why It Matters

The production LLM landscape in 2026 is not a landscape of single-prompt, single-response systems. The systems that engineering teams are building and operating are retrieval-augmented, tool-using, and increasingly agentic. That architectural shift has produced a new tier of production complexity that the patterns for operating a single LLM call do not address. A RAG system introduces retrieval quality as a production variable distinct from model quality and requires measurement, monitoring, and independent improvement. A tool-using system introduces action semantics: the outputs of these systems are not text on a screen but operations in the world, with real costs and real side effects. An agentic system compounds both concerns across multiple steps, with error propagation, cost accumulation, and state management problems that do not exist in single-step architectures.

The workflow-level impact of getting these patterns wrong is qualitatively different from getting a single-prompt system wrong. A language model that produces a suboptimal response loses one user interaction. An agent that loops on a booking API costs money, creates real-world side effects that must be reversed, and potentially exposes the organization to compliance or financial risk before any human notices something is wrong. The mission-aligned stance is to treat RAG, tool-calling, and agentic architectures as production engineering problems with distinct playbooks, not as model features that work automatically once the model is capable enough. Operational clarity here means that every system has a documented shape, and every shape has a matching set of production controls.

The maturity gap in most teams is not a capability gap but a discipline gap. The technical building blocks for RAG, tool-calling, and agentic systems are widely available; the frameworks, APIs, and models that enable them have proliferated faster than the operational practices that make them safe to run. Teams that build these systems by assembling capable components without applying production engineering discipline are carrying a debt that compounds with scale. The pilot-to-production pathway for any system in this chapter requires explicit answers to the questions: what is the retrieval quality contract, what happens when a tool call fails, what bounds the cost and step count of an agent run, and how does an operator reconstruct what happened during a session?

14.3 RAG, Tool-Calling, and Agentic: The Three Shapes of Modern LLM Apps

Distinguishing the architectural shape of an LLM feature is the first step in applying the right production playbook. Teams that conflate these shapes apply the wrong controls: a pure prompt system has no retrieval layer to monitor, but a RAG system does; a RAG system does not take actions in the world, but a tool-using system does; a tool-using system does not maintain state across multiple reasoning steps, but an agentic system does. The shape determines the failure surface, the cost profile, the observability requirements, and the safety controls that apply.

A pure-prompt system sends a context-enriched request to a language model and returns the response. The production concerns are latency, cost per token, output quality, and safety of the model's output. A retrieval-augmented generation system adds a retrieval step before the language model call: the user's query is used to retrieve relevant documents from a corpus, those documents are assembled into the prompt context, and the language model generates a response grounded in the retrieved material. The production concerns expand to include retrieval quality, corpus freshness, embedding model coherence, and the lineage connecting the response to the source documents.

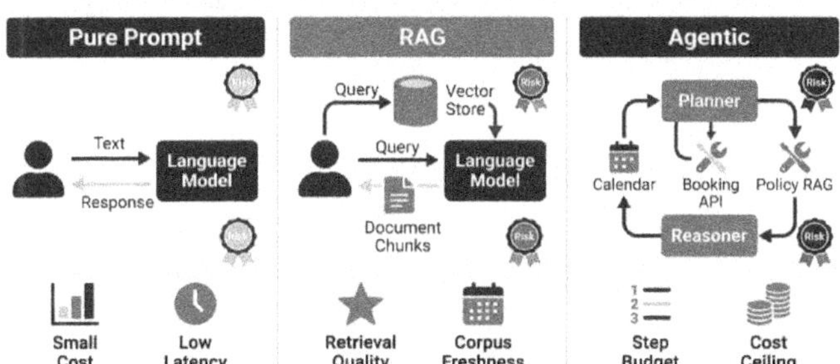

A tool-using system extends the RAG pattern by enabling the language model to invoke external tools, such as APIs, database queries, code execution environments, and communication systems. The model receives the results of tool calls as new context and can make additional tool calls or generate a final response. The production concerns include the idempotency of tool integrations, the costs and side effects of tool calls, and the error-propagation paths when tools fail. An agentic system is a tool-using system with an explicit reasoning loop: the model plans a sequence of steps, executes tools to gather information or take actions, observes the results, revises the plan, and repeats until the task is complete or a termination condition is reached. The production concerns are the cumulative concerns of all previous shapes plus state management, loop termination, cost ceilings, and multi-step observability.

The practical classification question for any LLM feature is: does it retrieve from a corpus, does it call external tools, and does it reason across multiple steps? The answer determines which sections of this chapter's playbook apply. A feature that retrieves from a corpus but does not call tools needs the retrieval quality, embedding pipeline, and corpus management patterns. A feature that calls tools needs the idempotency, scoped permissions, and tool schema patterns. A feature that reasons across multiple steps needs all of the above, plus the agent loop, cost ceiling, state management, and session trace patterns.

14.4 Part I — RAG in Production

The sections that follow form a self-contained arc on operating retrieval-augmented generation in production, and they are grouped deliberately because RAG is the most common shape a production LLM system takes. The next several sections cover the components that determine whether a RAG system works in practice: retrieval quality measured as recall, precision, and relevance; the operations of embedding models and vector indexes; the chunking and document-parsing decisions that govern what the retriever can find; and the hybrid search and re-ranking patterns that combine lexical and semantic signals into a usable result set. These are presented in sequence because they compose: a chunking decision constrains what embeddings can represent, which constrains what retrieval can recall, which constrains what re-ranking can promote.

The framing that organizes this arc is a deliberate inversion of how RAG is usually described. A RAG system is commonly introduced as a language model that happens to retrieve, with the model treated as the system and retrieval as a feature bolted on. In production, the relationship is the other way around. Operating a RAG system well is mostly a retrieval problem with a language model on top, because the dominant failure mode is not the model generating poorly but the retriever supplying the wrong context, and no amount of model capability recovers an answer grounded in irrelevant or stale material. The sections in this part therefore treat retrieval as the primary engineering surface and the model as the consumer of whatever retrieval delivers. A team that internalizes this ordering will instrument and budget its retrieval layer with the same seriousness it applies to model serving, which is the operational posture the rest of this part assumes throughout.

14.5 Retrieval Quality: Recall, Precision, and Relevance Patterns

Retrieval quality is the single most impactful variable in a RAG system, and it is the variable most likely to be treated as solved at launch and unmonitored in production. A language model with a poor retrieval context will produce poor answers regardless of its own capabilities. When those poor answers surface in product metrics, the investigation typically starts with the model rather than the retrieval layer, because the model is more visible. This misattribution cycle wastes engineering cycles and delays the actual fix. The production discipline is to instrument retrieval quality as a first-class signal from the first day of operation, alongside serving latency and error rate.

Recall at K is the primary offline metric for retrieval quality: of all the relevant documents for a given query in the evaluation set, how many appear in the top K retrieved results. Mean Reciprocal Rank measures where the first relevant result appears in the ranked list, penalizing systems that find relevant documents but bury them. The empty retrieval rate, the fraction of production queries for which the system returns zero results or all results below the minimum relevance threshold, is a coarse but operationally important signal that identifies query classes the current corpus cannot serve. All three metrics must be computed against a maintained evaluation dataset, updated as the corpus and query distributions evolve.

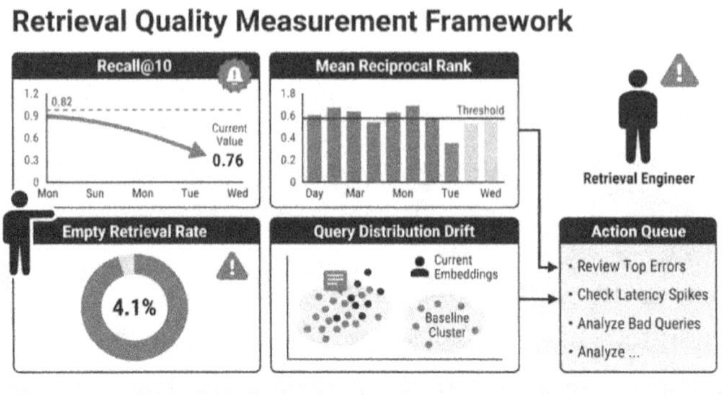

Retrieval Quality Measurement Framework

Online retrieval observability requires logging at the retrieval layer. Each retrieval event should record the query identifier, the retrieved document identifiers, the retrieval scores, and the per-stage latency. This event log is the data source for continuous quality measurement via sampled human evaluation, for query distribution drift detection, and for debugging individual retrieval failures. A retrieval log that captures only document identifiers and scores is sufficient for quality monitoring; a retrieval log that also captures the query embedding enables drift detection against the embedding distribution at indexing time, which surfaces the class of silent failure where the query distribution has shifted away from the corpus distribution without any change to either the model or the index.

Query expansion is a retrieval quality lever that improves recall for queries that are underspecified, use different vocabulary than the corpus, or describe a concept that the corpus discusses in different terms. A query expansion step rewrites the user's query into multiple alternative formulations, each retrieved independently, with the results merged before re-ranking. The expansion can be generated by a small language model or by rule-based synonym expansion from a domain vocabulary. The production cost is additional retrieval latency proportional to the number of expansions; the quality gain is most pronounced in corpora with a wide vocabulary gap between users and documents, such as technical documentation queried by non-technical users.

14.6 Embedding Models and Vector Index Operations

The embedding model is a transformation function that maps text to a vector space where semantic similarity is approximated by geometric proximity. Every downstream production decision depends on the quality of the mapping. Selecting an embedding model for a production RAG system is not primarily a benchmark question; it is a distribution question. The model must perform well on the specific vocabulary, domain, and query patterns of the production corpus and query distribution, not on a general-purpose benchmark. A model that ranks first on a public benchmark may rank third on an enterprise technical documentation corpus because the benchmark's query distribution differs from the production distribution. The selection process should include evaluation on a representative sample of production queries.

Dimension and index trade-offs are the first operational decisions after model selection. Higher-dimensional embeddings carry more information but increase memory consumption, index build time, and query latency. Many embedding models support matryoshka representation learning, which allows truncating the embedding to a lower dimension while incurring a controlled quality cost. The practical decision is to select the smallest dimension that meets the retrieval quality requirements for the production corpus, evaluated on the maintained quality dataset, rather than defaulting to the maximum dimension supported by the model. A team that defaults to maximum dimensions without evaluating quality at smaller dimensions incurs storage and latency costs that may not be justified.

Embedding Model Selection and Dimension Trade-off

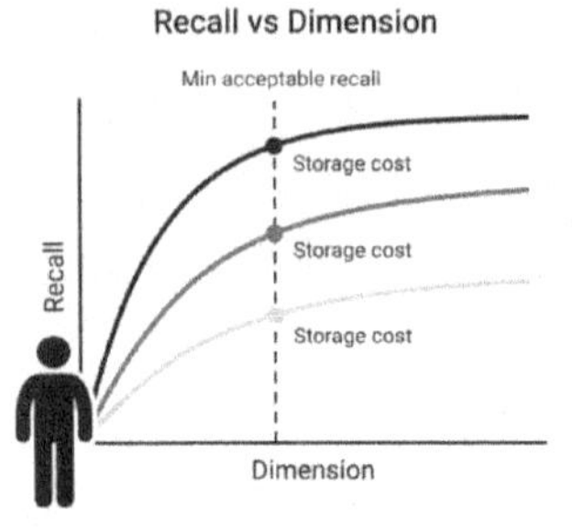

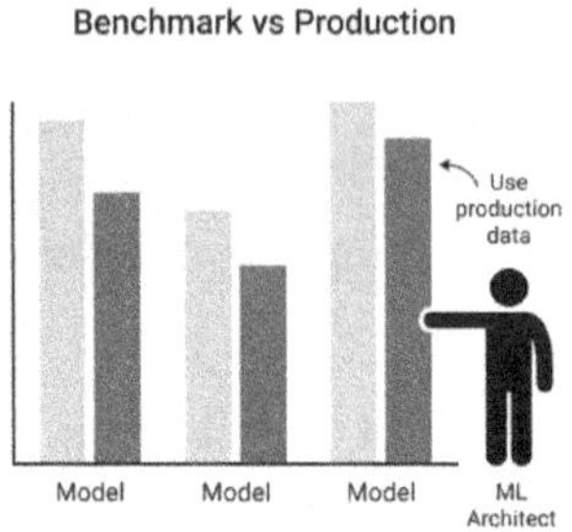

Vector index operations in production require planned maintenance that most teams do not budget for at launch. The index must support incremental insertion for new documents without requiring a full rebuild. It must support deletion for document lifecycle management and compliance requirements. It must support online queries during index updates, which requires either a lock-free design or a shadow index strategy where updates are applied to a staging index that is promoted to production after a validation check. The maintenance cadence, the expected index size at twelve and twenty-four months of growth, and the rebuild time for a full re-indexing event should all be documented in the system's operational runbook before launch.

Sharding a vector index is required when the corpus size exceeds the memory or throughput capacity of a single index. The sharding decision determines how queries fan out across shards, how shard results are merged, and how shard rebalancing is handled when the corpus grows unevenly across domains or document categories. Namespace-based sharding, where each shard holds a subset of the corpus defined by a metadata attribute such as document category or access group, enables efficient filtered retrieval by routing queries only to the relevant shards. Hash-based sharding distributes load more evenly but requires querying all shards for every request. The sharding strategy should be chosen based on the application's query filter patterns, not on general infrastructure defaults.

14.7 Chunking, Document Parsing, and Knowledge Curation

Chunking is the decision that most directly shapes retrieval quality, and it is a decision that most RAG systems make once, informally, and never revisit. A chunk is the unit of retrieval: the granularity at which documents are embedded, indexed, and returned as context to the language model. Chunks that are too large return irrelevant surrounding text alongside the relevant passage, degrading the signal-to-noise ratio of the context window. Chunks that are too small lose the surrounding context that makes a passage coherent, producing retrievals that are literally accurate but contextually incomplete. Neither failure is obvious in small-scale testing; both become visible at production scale when the corpus is large, and the query distribution is broad.

Document-structure-aware parsing extracts the semantic structure of a document before chunking: headings, subheadings, lists, tables, code blocks, and paragraphs are identified and preserved as structure rather than flattened into a continuous text stream. Structure-aware chunking then uses the document's own hierarchy as chunk boundaries, so that each chunk is a coherent structural unit rather than an arbitrary token window. For technical documentation, policy texts, and procedural documents where the structure is meaningful, structure-aware chunking produces significantly better retrieval than fixed-token chunking because the resulting chunks align with the granularity at which users actually ask questions.

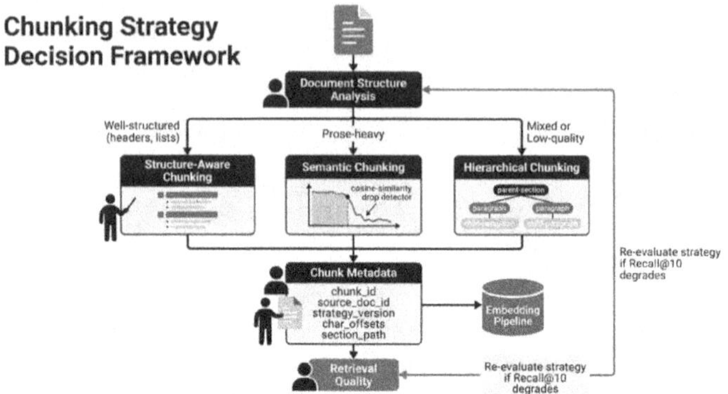

Knowledge curation is the ongoing practice of maintaining the quality, freshness, and coherence of the document corpus. A corpus that only grows will eventually contaminate retrieval with stale, duplicate, or low-quality content. Curation requires active processes: near-duplicate detection to prevent multiple versions of the same document from inflating retrieval scores for a single topic, document lifecycle tracking to identify and retire deprecated content, quality scoring at ingestion time to flag documents that are too short, too dense with numbers and formatting, or in a language outside the supported set. The curation infrastructure is not glamorous engineering; it is the invisible discipline that determines whether the retrieval layer improves or degrades over time.

The chunking strategy is a versioned artifact. The optimal strategy for a corpus at launch may not be optimal at six months when the corpus has grown, the query distribution has shifted, or the embedding model has been updated. The strategy version, the evaluation dataset it was optimized against, and the quality metrics it achieved should be recorded alongside the embedding model version in the chunk metadata. When an embedding model upgrade triggers a re-indexing event, the team should also evaluate whether the chunking strategy should be updated, treating the re-indexing as an opportunity to improve both dimensions simultaneously, rather than running a re-indexing that holds the chunking strategy constant.

14.8 Hybrid Search, Re-Ranking, and Fusion Patterns

Dense retrieval uses embedding similarity as its ranking signal. It generalizes well across semantic similarity, paraphrasing, and conceptual abstraction. Still, it performs poorly on queries that contain rare terms, proper nouns, product codes, internal identifiers, or exact phrases that are not well-represented in the embedding model's training distribution. Sparse retrieval, implemented as BM25 or its variants, inverts this failure: it performs well on exact-term matches but poorly on queries that rely on semantic understanding across vocabulary differences. The production conclusion is that neither approach alone is optimal across the full distribution of queries a real system receives. Hybrid retrieval, which combines both signals, is the production default, not a sophisticated extension reserved for high-priority applications.

Reciprocal Rank Fusion is the most operationally robust method for combining ranked lists from dense and sparse retrieval. Each document receives a score proportional to the reciprocal of its rank in each list, and those scores are summed. The method requires no score normalization, no calibration against a held-out query set, and no hyperparameter tuning beyond a small constant that sets the floor rank. It is robust to the differences in score scale and distribution between dense and sparse systems. More complex learned fusion methods can achieve higher quality on a stable, well-characterized query distribution. Still, their quality advantage is typically small, and their calibration cost and distribution-shift fragility make them a suboptimal choice for most production deployments.

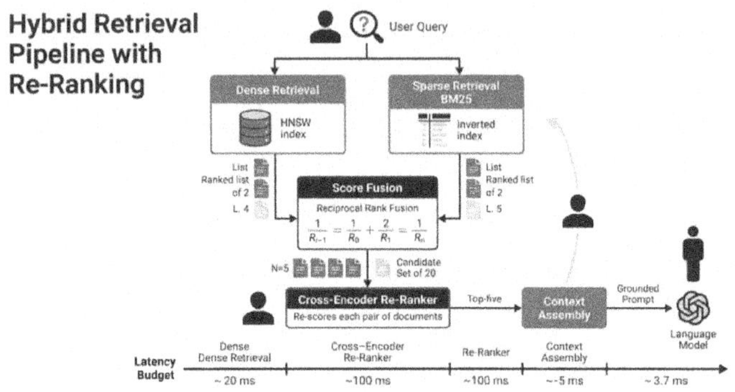

Cross-encoder re-ranking is the final quality lever in the retrieval pipeline. A cross-encoder processes each query-document pair jointly, producing a relevance score that is more accurate than bi-encoder embedding similarity because it can attend to the interaction between the query and the document rather than scoring them independently. The latency cost is high: a cross-encoder cannot precompute document representations, so it must process the full pair at query time. The production pattern is to re-rank a candidate set of 20 to 50 documents from the fusion step, not the full retrieval pool. This bounds the latency impact while capturing most of the quality improvement, since the fusion step has already eliminated most irrelevant documents.

Multi-source retrieval, in which the RAG system queries multiple document collections with different schemas, vocabularies, or access controls, requires a fusion step to handle heterogeneous result sets. Each source may have a different score scale, a different document granularity, and different metadata fields. The fusion step must normalize across these differences before merging. Source weighting, where results from more authoritative or higher-quality sources are boosted relative to lower-quality sources, is a fusion parameter that should be exposed as a configurable system setting rather than being hardcoded, because the relative authority of different sources changes as the corpus evolves and new sources are added.

14.9 Part II — Agents and Tool-Using LLMs in Production

The second arc of this chapter turns from retrieval to action. The sections that follow cover tool calling and function-calling patterns; the agent loop architectures that structure multi-step reasoning, including ReAct, plan-execute, and hierarchical designs; multi-agent orchestration; the step budgets and termination conditions that bound an agent run; the memory and state management that long-running agents require; the observability practices that make an agent trace legible; and the safety, sandboxing, and permissioning controls that constrain what an agent is allowed to do. These topics are grouped because they share a defining property that the retrieval arc does not: the outputs are operations in the world, with real costs and real side effects, rather than text on a screen.

The framing for this arc is that agentic systems are an extension of LLM serving, not a separate discipline. An agent is still a model behind a serving stack, subject to the same latency, cost, and reliability concerns the earlier serving chapters established, and a team that has not mastered single-call serving will not master agents. What changes is the failure surface, which multiplies in ways the serving chapters do not address. A single model call fails in one place and is retried or returned as an error; an agent run fails across an unbounded sequence of steps, each of which can call tools, accumulate cost, propagate an earlier error, and take an irreversible action before any human notices. The sections in this part build on the serving foundations rather than replacing them, and they concentrate on the controls that contain a failure surface the earlier chapters never had to bound.

14.10 Tool Calling and Function-Calling Patterns

Function calling is the mechanism by which a language model signals its intent to invoke an external tool: the model receives a set of tool schemas describing available functions, their parameters, and their return types, and generates a structured call rather than free-text output when it determines that a tool should be invoked. The calling convention is a contract between the model and the integration layer. Ambiguous, underspecified, or inconsistently structured tool schemas produce ambiguous tool calls; the integration layer must validate the call structure and handle cases where the model generates a syntactically valid but semantically incorrect call.

Idempotency is the non-negotiable requirement for any tool integration that takes an action in the world. An idempotent tool call produces the same result when called multiple times with the same parameters as when called once. This property matters because agents retry failed calls, because network failures can cause calls to be retried at the infrastructure level, and because agent loops can generate duplicate calls through the same ambiguity-and-retry cycle that produced forty-three booking API calls in the opening scenario. A tool integration that is not idempotent must provide a compensating action that reverses the side effects of a duplicate call, and that compensating action must be documented in the tool schema and tested as part of the integration.

Tool Integration Patterns and Idempotency

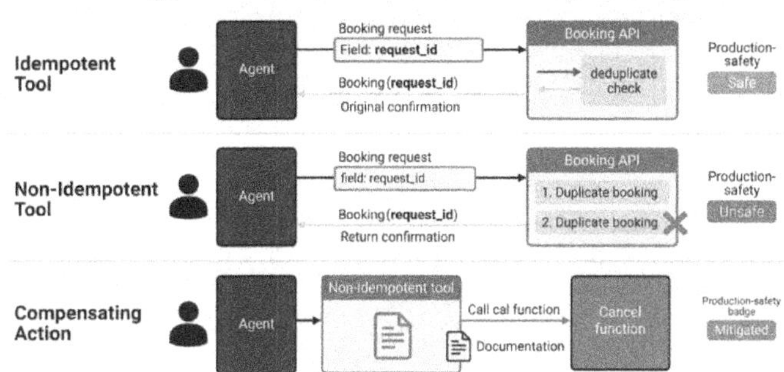

Tool schema design determines the quality of the model's tool-calling behavior. A well-designed schema provides enough information for the model to call the tool correctly in the common case, explicitly handles edge cases through parameter descriptions and enum constraints, and makes the error response structure as informative as the success response structure. Error responses that return an opaque error code provide the model with no information about how to recover; error responses that explain what was wrong and what the caller should do differently enable the model to adapt its next call. The tool schema is a user-interface design problem where the user is the language model, and clarity in the schema directly translates to accuracy in the calls.

Permissioning for tool-using systems must be scoped to the minimum access required by each tool. A tool that reads calendar events should not hold credentials that also allow writing calendar events. A tool that queries a database should have read-only credentials scoped to the tables it actually needs. Broad credentials held by any component of a tool-using system represent an expanded blast radius. If the agent can be manipulated into calling a tool in an unintended way, the scope of the damage is bounded by the permissions of that tool's credentials. Scoped credentials are a safety control, not an infrastructure convenience, and they should be reviewed as part of any security assessment of a tool-using system.

14.11 Agent Loop Patterns: ReAct, Plan-Execute, and Hierarchical

The ReAct pattern interleaves reasoning and action in a single continuous loop: the agent produces a thought that explains its current understanding and intent, emits a tool call or a final answer, observes the result, and repeats. The loop continues until the agent produces a final answer or a termination condition is reached. ReAct agents are the simplest agentic architecture and the most widely deployed. Their failure modes are characteristic: the agent can get stuck in a reasoning loop where each thought concludes that more information is needed, leading to cascading tool calls; the agent can misinterpret a tool response and proceed with incorrect information; and without an explicit step budget, the agent can consume unbounded tokens and cost across arbitrarily many reasoning steps.

Plan-and-execute architectures separate planning from execution. A planner model receives the task and produces a structured plan: an ordered list of steps, each with a specified tool call, the expected input, and the expected output. An executor model then runs each step in sequence, passing the outputs of earlier steps as inputs to later steps. The plan is validated before execution begins, which surfaces ambiguities and impossible steps before any tool calls are made. The failure mode of plan-and-execute is plan brittleness: a plan that was valid when produced may become invalid when an early step produces an unexpected output. Robust plan-and-execute implementations include a plan revision step that can modify the remaining plan in response to unexpected intermediate results.

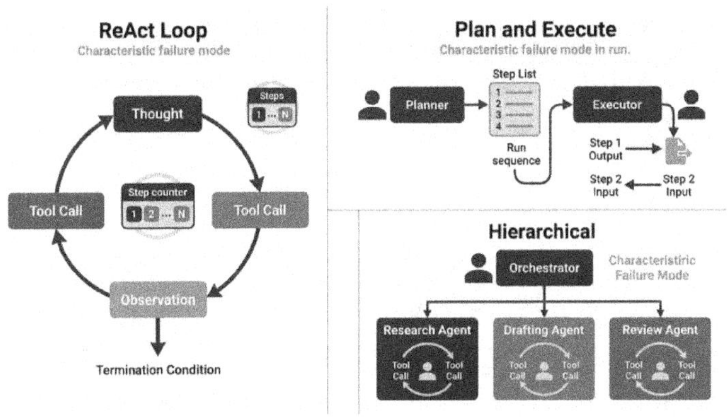

Hierarchical agent architectures introduce an orchestrator that decomposes a complex task into subtasks and delegates each subtask to a specialist agent. The orchestrator is responsible for task decomposition, delegation, result aggregation, and inter-agent coordination. Specialist agents are responsible for executing their assigned subtask within their area of capability. The production benefits of hierarchical architectures include scope containment (each specialist operates within a bounded context with a smaller tool set and a simpler task) and parallelism (independent subtasks can be executed concurrently). The production risks are coordination complexity (the orchestrator must correctly decompose the task and interpret each specialist's results) and cascade failure (if one specialist produces an incorrect result, the orchestrator may propagate that error into the final output).

The characteristic failure mode of each loop pattern is worth naming explicitly because teams tend to discover it in production rather than in testing. ReAct agents loop on ambiguity. Plan-and-execute agents fail when the plan assumes a world state that the environment does not match. Hierarchical agents fail when the task decomposition is incorrect or when the orchestrator aggregates specialist results that are mutually inconsistent. A more capable model can eliminate none of these failure modes; they can only be mitigated by explicit loop-termination logic, plan validation, and result-consistency checks built into the architecture.

14.12 Multi-Agent Systems and Orchestration Patterns

Multi-agent systems are architectures in which multiple autonomous agents collaborate to complete tasks beyond the scope or capabilities of a single agent. The collaboration may be sequential, where one agent's output is the next agent's input; parallel, where multiple agents work on independent subtasks simultaneously; or iterative, where agents review and revise each other's work across multiple passes. Each coordination pattern has a different operational profile: sequential pipelines are simple to trace but create single points of failure at each handoff; parallel systems reduce latency but require a result aggregation step that handles conflicting or inconsistent outputs; iterative systems can improve quality through revision but require a convergence criterion to prevent infinite refinement loops.

Explicit orchestrators provide operational clarity that implicit coordination does not. An explicit orchestrator is a designated component responsible for task decomposition, agent dispatching, result collection, and final aggregation. Its state machine is visible and auditable: at any point, an operator can inspect the orchestrator's current state to determine which agents are active, which subtasks are complete, and what inputs each active agent is processing. Implicit coordination, where agents communicate through a shared message bus without a designated orchestrator, is more flexible but harder to trace, debug, and bind in terms of cost and step count.

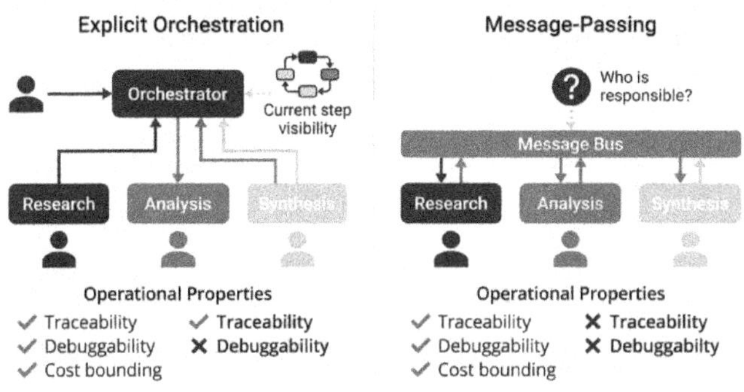

Message-passing architectures, where agents communicate by publishing and consuming messages on a shared bus, offer a flexibility that explicit orchestration does not: any agent can respond to any message, agents can be added or removed without changing the coordination logic, and the architecture scales horizontally with the message volume. The production cost of this flexibility is that the coordination logic is distributed and implicit: no single component has a complete view of the system's state, tracing the path from a task submission to a final result requires reconstructing a sequence of events from the message log, and preventing coordination loops requires each agent to implement its own loop detection and termination logic. The pattern is appropriate for systems with open-ended, dynamic task spaces; it is a poor choice for systems where auditability, cost bounding, and predictable behavior are primary requirements.

Agent spawning policies govern when an orchestrator may create new agents. Without a spawning policy, an orchestrator facing a complex task may recursively decompose it into sub-tasks and spawn new agents for each, producing an agent tree that grows exponentially. The spawning policy sets limits on the maximum number of concurrent agents, the maximum depth of agent delegation, and the total budget across all agents in a session.

These limits are not a constraint on capability; they are production controls that make multi-agent systems safe to operate on customer-facing traffic, where the cost and latency implications of unbounded spawning are operationally unacceptable.

14.13 Cost Ceilings, Step Budgets, and Termination Conditions

The cost ceiling is the hardest operational control for an agentic system. Without a cost ceiling, an agent running on a complex or ambiguous task will consume tokens until it either completes the task or is stopped by an external intervention. At scale, across hundreds of concurrent agent sessions, the absence of cost ceilings produces a cost surface that is unbounded and unforecastable. A cost ceiling is a per-session limit on total token consumption, measured in either raw token count or monetary cost, depending on the serving infrastructure's billing model. When the ceiling is reached, the session is terminated cleanly: the agent produces a partial result with an explicit indication that it was terminated due to budget exhaustion, rather than failing silently or producing a response that appears complete but is truncated mid-reasoning.

Step budgets are the complementary control to cost ceilings. A cost ceiling bounds the total resource consumption of a session; a step budget bounds the number of reasoning cycles, or tool calls, the agent is permitted to execute. Step budgets are the more direct means of preventing reasoning loops: a step budget of twenty ensures that an agent that enters a reasoning loop will be terminated after twenty iterations, regardless of the per-step token cost. The two controls are not redundant: a step budget without a cost ceiling can be violated by an agent that produces very long thoughts on each step; a cost ceiling without a step budget can be violated by an agent that executes many cheap steps in a tight loop.

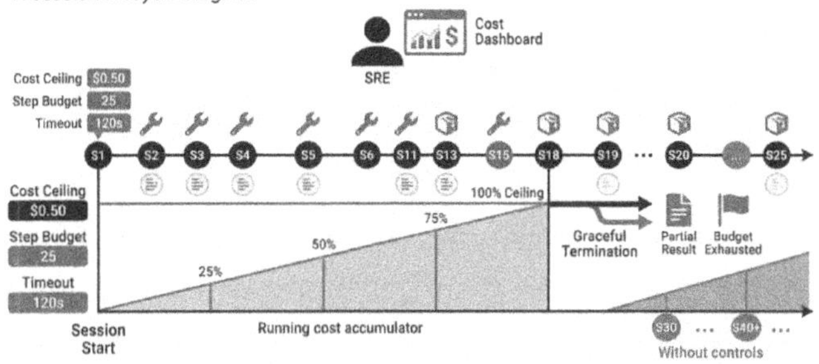

Termination conditions are the explicit logic that determines when an agent session has reached a valid conclusion. A session terminates successfully when the agent produces a final answer that passes the completion criteria; it terminates with a budget exhaustion signal when the cost ceiling or step budget is reached; it terminates with an error when the agent produces an output that fails a safety check, calls a tool that returns an unrecoverable error, or enters a state that the termination logic has classified as a dead end. Each termination path must produce a session record that captures the terminal state and the reason for termination, because the distribution of termination reasons is the primary signal for diagnosing systemic agent behavior problems.

Graceful degradation under budget pressure is an architectural decision that most teams make implicitly rather than explicitly. When a cost ceiling is approaching, the agent can continue operating until the ceiling is hit and then terminate, or it can detect the approaching ceiling and switch to a lower-cost completion strategy: producing a partial answer based on what has been gathered so far, or asking the user for clarification rather than continuing to gather information autonomously. The right strategy depends on the application context. For customer-facing applications, a partial answer with a clear explanation is almost always preferable to a silent truncation or an abrupt error. This preference should be encoded in the agent's termination logic rather than left as an emergent behavior of the model.

14.14 Memory, State, and Long-Running Agent Patterns

Agent memory is the mechanism by which information persists across the steps of a single session and, for long-running or multi-session agents, across sessions. Short-term memory is the agent's context window: the working memory of the current session, containing the conversation history, tool call results, and reasoning trace up to the current step. Long-term memory is an external store that the agent can read from and write to across sessions: a structured database of facts, a vector store of past interactions, or a document store of prior task outputs. The operational challenge of long-term memory is all the challenges of any production data store, compounded by the fact that the agent is both a consumer and a producer of the data it depends on.

Context window management is the primary operational challenge of short-term memory. As a session progresses, the accumulated context grows: the initial system prompt, the conversation history, the retrieval results from each RAG call, the tool call inputs and outputs, and the reasoning traces all compete for the finite context window. A context that exceeds the model's window limit truncates the earliest content, potentially removing information critical to the current step that was generated early in the session. The production patterns for context window management include rolling summaries (replacing older conversation turns with a compressed summary), selective retention (keeping only the most relevant tool outputs rather than the full history), and explicit context pruning triggered when the context approaches the window limit.

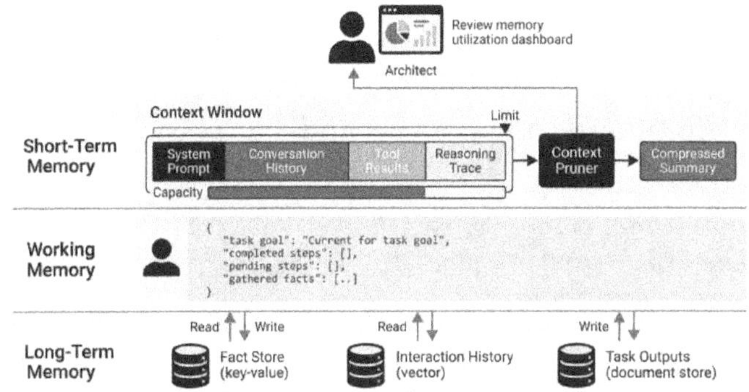

Long-term memory introduces state corruption risks that compound over time. An agent that writes incorrect information into its long-term memory stores, or that accumulates outdated information from prior sessions, will have its future behavior influenced by those errors. The longer the agent operates, the more opportunities there are for incorrect writes to propagate. Production-safe long-term memory requires write policies that distinguish high-confidence facts from speculative information, expiry policies that retire stored information that has exceeded its expected validity window, and audit logs that record what the agent wrote to long-term memory and under what conditions, thereby enabling retroactive identification and correction of incorrect stored facts.

Checkpointing is the operational mechanism that enables long-running agents to be resumable. An agent session that spans hours or days, or that requires human approval at intermediate steps, must be able to suspend and resume without losing the reasoning state accumulated up to the point of suspension. A checkpoint captures the full agent state: the current context window, the working memory, the list of completed and pending steps, and the outputs gathered so far. Resuming from a checkpoint restores this state and continues execution from where it was suspended. The checkpoint store must be durable and consistent; a lost or corrupted checkpoint leaves the agent in a state where resumption is impossible, and the session must be restarted from the beginning.

14.15 Observability for RAG and Agentic Systems

Observability in multi-step LLM systems is structurally different from that in single-call LLM systems. A single LLM call produces one request, one response, one latency measurement, and one cost measurement. An agentic system produces a session: a sequence of reasoning steps, tool calls, retrieval events, and model calls, each with its own latency, cost, and output. The session is the unit of observability for agentic systems, and the session trace is the data structure that enables understanding, debugging, and improvement of what happened during a session. A session trace that captures only the agent's final input and output is as useful for debugging as a stack trace that shows only the top and bottom frames.

Session traces should be structured rather than free-text. Each step in the trace is a structured event: step index, step type (reasoning, tool call, retrieval, final answer), inputs, outputs, latency, token count, and cost. The trace events are linked by session identifier, enabling reconstruction of the complete session from the event log. Structured traces support programmatic analysis: identifying sessions that exceeded the step budget, identifying the tool calls that contributed most to session cost, detecting sessions in which the same tool was called multiple times with identical parameters, and measuring the distribution of session lengths and termination reasons across the production workload.

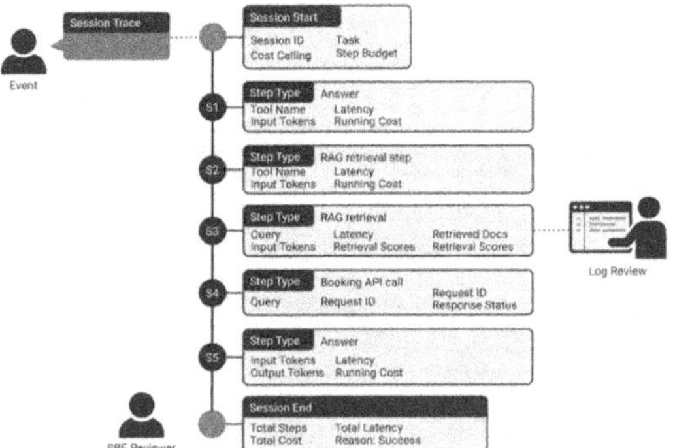

Retrieval-quality observability within an agentic system requires instrumenting the retrieval layer at the per-query level and associating each retrieval event with the session trace. An agent that executes multiple retrieval queries during a session should produce a retrieval event for each query, linked to the session by session identifier and to the step trace by step index. This linkage enables root-cause analysis that distinguishes agent failures caused by poor retrieval from agent failures caused by poor reasoning: a session that terminates with an incorrect final answer can be traced back to the specific retrieval queries that preceded the incorrect reasoning step, and the retrieval scores for those queries can be inspected to determine whether the failure was a retrieval quality problem or a reasoning problem.

Alerting for agentic workloads must extend to session-level signals. The rate of sessions exceeding the step budget, the fraction of sessions terminated by the cost ceiling rather than by successful completion, the distribution of termination reasons, and the mean session cost are the primary operational signals indicating whether the agent architecture is performing within its designed parameters. A sudden increase in the fraction of sessions that hit the cost ceiling may indicate a model behavior change, a change in the distribution of incoming tasks, or a tool integration failure causing the agent to retry tool calls repeatedly. These signals must be surfaced in the operational monitoring layer and owned by an on-call rotation, not discovered through downstream product metric declines.

14.16 Safety, Sandboxing, and Permissioning for Tool-Using Agents

The safety posture of a tool-using agent is determined by the union of the capabilities of all its tools. An agent with a tool that sends email, a tool that reads the calendar, and a tool that calls an external API has, by construction, the ability to send emails to any recipient, expose calendar data to the API, and chain these operations in sequences that may not have been anticipated in the safety review. The attack surface is the tool set, and the safety analysis must consider not only the intended use of each tool but also the potential for tool chaining that produces unintended effects. This analysis is more demanding than reviewing the safety of any individual tool because tool chaining creates emergent capabilities that are not present in any individual tool.

Sandboxing is the technical control that limits the blast radius of a tool call gone wrong. A sandboxed execution environment for code execution tools runs untrusted code in an isolated container with no network access, no file system access outside a designated scratch directory, and no access to the host system's resources. A sandboxed API integration connects to external services through a proxy that enforces rate limits, logs all calls, and blocks requests that match abuse patterns. Sandboxing does not prevent all misuse; it bounds the damage that misuse can cause and makes misuse detectable. Both properties matter for production safety.

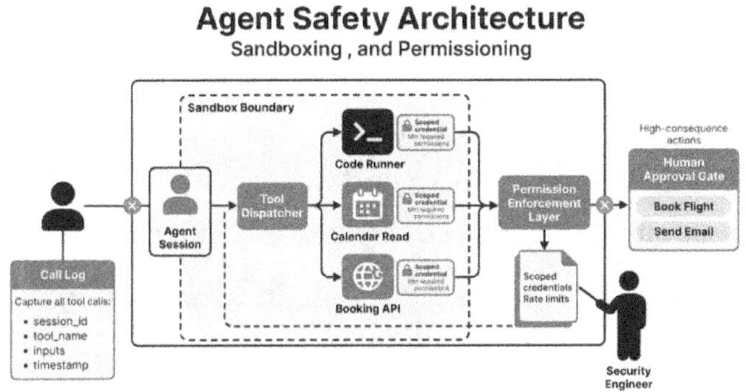

Human-in-the-loop confirmation gates are the safety control for high-consequence, irreversible tool calls. A booking, a payment, a deletion, a message sent to an external party, or any other action whose reversal is costly or impossible should require explicit human confirmation before execution, at least during the initial deployment period when the agent's behavior on the full production distribution of tasks has not yet been characterized. The confirmation gate should present the human with a clear description of the action to take and its expected effects, not with the raw tool-call parameters. An operator who must approve a booking API call in the form of a JSON payload with a dozen fields is not in a position to make an informed approval decision; an operator who sees a natural-language summary of the booking is.

Prompt injection is the security concern specific to tool-using and agentic systems that interleave model outputs with content retrieved from external sources. When a retrieved document contains text crafted to appear like an instruction to the model, the model may execute that instruction rather than treating the document as context. A retrieval result that instructs the agent to ignore previous instructions and call a specific tool, or to forward gathered information to an external endpoint, constitutes a prompt-injection attempt. The defenses are a combination of architectural separation (never placing retrieved content in the system prompt or in positions the model treats as authoritative instructions), output validation (checking agent outputs for patterns associated with injection-induced behavior), and sandboxing (limiting the blast radius of a successful injection to the permissions of the compromised tool).

14.17 Production RAG Operations: Index Refresh, Drift, and Quality Degradation

A RAG system that retrieves correctly on launch day is not a RAG system that retrieves correctly six months later, and the gap between those two states is the discipline of RAG operations. RAG implementation is the work of building the pipeline: choosing an embedding model, configuring a vector index, defining a chunking strategy, and assembling retrieved context into a prompt. RAG operations is the work of keeping that pipeline correct as the corpus grows, documents are deleted, the query distribution shifts, and the embedding model is eventually replaced. The two are distinct because the failure modes are distinct. An implementation failure produces a system that never worked; an operations failure produces a system that worked, was declared done, and degraded silently while the team's attention moved elsewhere. The sections of this part have treated retrieval as the primary engineering surface of a RAG system, and that framing carries a direct operational consequence: a retrieval surface that is not refreshed, monitored, and budgeted will degrade on the same timescale as any other unmaintained production component.

Index refresh is the first operational concern, and its cadence is a deliberate decision rather than a default. A corpus that changes daily, such as a support knowledge base or a product catalog, requires a refresh pipeline that ingests new and changed documents on a schedule short enough that the retrieval layer reflects current reality. The refresh can be triggered on a fixed interval, on a volume threshold of accumulated changes, or on an event from the upstream system of record, and the right trigger depends on how quickly stale content produces wrong answers. The operational requirement is that refresh is a named, monitored pipeline with its own success and failure signals, not an unobserved background job. A refresh pipeline that fails silently for a week produces a retrieval layer that confidently returns last week's reality, and the only symptom visible downstream is a gradual rise in answer-quality complaints that the team will initially misattribute to the model.

Document deletion is the operational concern most implementations handle incorrectly, because deletion in a vector store is not the simple operation it is in a relational database. When a source document is removed or superseded, its embeddings must be removed from the index, or the retriever will continue to surface content that no longer exists in the system of record. Many vector stores implement deletion as a tombstone: the vector is marked deleted and excluded from results but not immediately reclaimed from the index, which preserves query performance at the cost of index size. The operational requirement is a deletion path that is as reliable as the ingestion path, with tombstoning understood as the mechanism rather than assumed to be immediate physical removal, and a periodic compaction that reclaims tombstoned space before it degrades index performance. A RAG system that ingests reliably but deletes unreliably will accumulate retrievable content that contradicts the current corpus, which is among the hardest RAG failures to diagnose because the retrieved content looks legitimate.

Embedding model migration is the highest-risk routine operation a RAG system undergoes, because the embedding model defines the geometry of the entire index. When a newer embedding model offers better retrieval quality, adopting it requires re-embedding the full corpus, since vectors produced by two different embedding models are not comparable and cannot coexist in a single similarity search. The migration must be performed without service disruption, which means building the new index alongside the existing one, validating retrieval quality on the new index against a held-out evaluation set before any traffic moves, and cutting over only after the new index demonstrates equal or better quality. The cutover itself should be staged, routing a fraction of queries to the new index and comparing retrieval quality before completing the migration. A team that re-embeds in place, replacing vectors as the new model produces them, will run a window during which the index contains an incoherent mixture of two geometries, and retrieval quality during that window is undefined.

Recall-precision drift detection is the monitoring discipline that makes the preceding concerns observable. Retrieval quality degrades through two distinct mechanisms, and a production system should detect both. Recall drift occurs when relevant documents stop appearing in the retrieved set, often because the corpus grew and the relevant content is now crowded out by near-neighbors, or because a chunking change altered what the retriever can find. Precision drift occurs when irrelevant documents increasingly appear in the retrieved set, diluting the context the model receives. Detecting either requires a maintained evaluation set of representative queries with known-relevant documents, scored on a schedule against production retrieval, because retrieval quality is not visible in serving metrics like latency and error rate. A RAG system instrumented only for serving health is blind to the degradation that matters most, and the team will learn about it from users rather than from monitoring.

The stale-retrieval failure mode is the canonical RAG operations incident and deserves naming because it is so frequently misdiagnosed. Stale retrieval occurs when the retrieval layer returns content that was correct at index time but is no longer correct, because the source changed and the index did not. The symptom is a system that answers confidently and incorrectly, grounded in real source material that has since been superseded, which makes the failure especially convincing and especially hard to detect through output inspection alone. Stale retrieval is an operations failure, not a model failure, and a team that responds by tuning the model or the prompt is treating a symptom. The correct response is to verify the refresh and deletion pipelines, confirm the index reflects the current system of record, and add freshness monitoring that compares index state against source state. A RAG system without freshness monitoring will experience stale retrieval and will spend its first occurrence debugging the wrong layer.

A/B testing retrievers behind a fixed model is the disciplined way to improve retrieval without conflating variables. Because retrieval quality and model quality are independent contributors to answer quality, the only way to know whether a retrieval change helped is to hold the model constant and compare two retrievers on the same query distribution. This requires routing traffic across two retrieval configurations, capturing the retrieved context and the resulting answer quality for each, and comparing them on the maintained evaluation set and on production signals. The same discipline applies to chunking changes, re-ranking changes, and hybrid-search weighting changes, each of which is a retrieval-layer intervention whose effect is measurable only when the model is held fixed. A team that changes the retriever and the prompt in the same release loses the ability to attribute the result to either.

Logging retrieval traces for replay is the operational practice that makes all of the preceding diagnosable after the fact. Every retrieval operation should log the query, the retrieved document identifiers, their relevance scores, and the final context assembled for the model, retained for a window sufficient to investigate a reported failure. With retrieval traces, a quality complaint becomes a replayable investigation: the team can reconstruct exactly what the retriever returned for the failing query, determine whether the failure was recall, precision, staleness, or a model error given correct context, and fix the responsible layer. Without retrieval traces, the same investigation is guesswork. Retrieval trace logging is the RAG equivalent of distributed tracing for a serving stack, and a RAG system operated without it is operated blind.

Cost-per-query budgets for retrieval close the operational frame, because retrieval is not free and its cost scales with the patterns a team adopts to improve quality. Each query may issue an embedding call, one or more vector searches, a re-ranking pass that can itself be a model call, and the assembly of a large context that the downstream model is then billed to process. A retrieval configuration tuned purely for quality can multiply the cost and latency of a query without the team noticing, because the cost is distributed across components rather than concentrated in the model call. The operational requirement is a per-query retrieval budget, measured and monitored, so that a quality improvement that doubles retrieval cost is a visible and deliberate trade-off rather than a silent regression in unit economics.

The operational maturity signals for production RAG are simply stated: retrieval is named, monitored, and budgeted. A mature RAG operation has named pipelines for refresh and deletion with their own health signals, monitored retrieval quality measured continuously against a maintained evaluation set rather than assumed from launch, and an explicit cost-per-query budget for the retrieval layer that finance recognizes. A RAG system that has all three has moved retrieval from an implementation detail to an operated production component, which is the difference between a RAG system that works in a demo and one that remains correct and economical as its corpus, its queries, and its embedding models evolve over the years it runs.

14.18 Production Agent Operations: Step Accounting, Cost Caps, and Failure Recovery

Of the three shapes this chapter treats, the agentic shape is the hardest to operate, because it is the only one in which a single user request expands into an unbounded sequence of model calls and tool invocations whose cost, duration, and side effects are not known in advance. A pure-prompt system has one call with a knowable cost. A RAG system adds a bounded set of retrieval operations before that call. An agent has a loop, and a loop is the operational difference between a system whose worst case is a single bad answer and a system whose worst case is a runaway sequence of real-world actions that accumulates cost and consequence until something external stops it. Agent operations is the discipline of bounding, recording, and recovering that loop, and it is the most consequential operational territory in this chapter because the failure modes it addresses are the ones that produce financial and operational incidents rather than merely degraded answers.

Per-trajectory step accounting is the foundational practice, and it begins with treating the trajectory as the unit of operation rather than the individual model call. A trajectory is the full sequence of steps an agent takes in service of one request: each model call, each tool invocation, each observation, and each plan revision, from the initial request to the terminal state. Step accounting records this sequence as an ordered, auditable log in which every step is attributable, timestamped, and costed. The auditable agent log is not optional instrumentation; it is the artifact that makes an agent operable, because every subsequent operational concern, including cost control, failure recovery, and accountability, depends on being able to reconstruct exactly what the agent did and in what order. An agent operated without per-trajectory step accounting cannot be debugged, cannot be audited, and cannot be safely bounded, because the system has no record of the sequence it is supposed to control.

Cost caps are the primary bound on a trajectory, and a mature agent system implements both hard and soft caps at two scopes. A soft cap is a threshold that triggers a warning, a downgrade to a cheaper model, or a request for human confirmation, allowing the trajectory to continue under altered conditions. A hard cap is a ceiling that terminates the trajectory when reached, regardless of completion state, because the alternative is an unbounded cost.

Caps are enforced per session, which bounds the cost a single user or conversation can accumulate, and per trajectory, which bounds the cost of a single request within that session. The two scopes are distinct because a session may legitimately contain many cheap trajectories or one expensive one, and a system that caps only at the session level cannot prevent a single runaway request from consuming the entire session budget in one trajectory. Cost caps are the control that converts the agent loop from an unbounded liability into a bounded operational risk.

Partial-completion semantics are the question a cost cap forces, and a system that has not answered it before deployment will answer it badly under load. When a hard cap terminates a trajectory mid-task, the agent has done some work and not all of it, and the system must define what that means. The two failure modes to avoid are leaving the world in an inconsistent intermediate state and reporting success when the task was abandoned. The operational requirement is that a capped trajectory returns a defined partial-completion result that records what was accomplished, what was not, and what state the external systems were left in, so that a human or a subsequent trajectory can resume or reverse from a known point. Partial completion is not an edge case to be handled later; it is the guaranteed outcome whenever a cap binds, and an agent system that treats it as exceptional will produce silent partial successes that are discovered only when their downstream consequences surface.

Idempotency requirements for tool calls are what make recovery possible, and they are a precondition rather than an enhancement. Because an agent may retry a step after a failure, resume a trajectory after an interruption, or repeat an action due to a planning error, every tool the agent can invoke should be idempotent or should be wrapped in a compensating mechanism that makes repeated invocation safe. A tool that charges a payment, sends a message, or creates a record is not naturally idempotent, and invoking it twice produces two charges, two messages, or two records. The operational requirement is that non-idempotent tools are either made idempotent through client-supplied idempotency keys or are guarded by a deduplication layer that recognizes a repeated invocation, with the alternative being an agent that double-acts whenever a retry or resume occurs. Idempotency is the property that lets the recovery patterns below operate without multiplying side effects.

Retry and resume patterns for multi-step trajectories follow directly from step accounting and idempotency. When a trajectory is interrupted, by a transient tool failure, an infrastructure event, or a deliberate pause for human confirmation, the system should be able to resume from the last recorded step rather than restarting from the beginning, which would repeat completed work and waste the accumulated cost. Resuming requires that the trajectory state, including the plan, the completed steps, and their results, is persisted at each step rather than held only in memory, so that the trajectory can be reconstructed and continued. Retry operates at the step level, reattempting a failed tool call under a bounded retry policy, while resume operates at the trajectory level, continuing a persisted trajectory after an interruption. Both depend on the auditable log as the source of truth for what has already been done.

Circuit-breaker patterns protect the agent from a degraded external dependency, applying a well-established serving pattern to the tool layer. When an external tool begins to fail or to respond slowly, an agent without a circuit breaker will continue invoking it, consuming steps and cost on a dependency that cannot serve, and potentially looping as it retries the failing tool. A circuit breaker monitors the failure rate of each tool and, when a threshold is crossed, opens the circuit so that further invocations fail fast rather than hanging or retrying, allowing the agent to take an alternative path or terminate cleanly. The circuit breaker is per tool, because tools degrade independently, and its state should be visible in the trajectory log so that a trajectory terminated by an open circuit is distinguishable from one terminated by a cost cap or a planning failure.

The agent-as-deployable-artifact frame ties the operational practices into the release discipline the rest of this book establishes for models and prompts. An agent is not a piece of code that calls a model; it is a composite artifact whose behavior is determined by its prompts, its available tools, and its operating policies, and each of those must be versioned and deployed as a unit. The versioned agent prompts define how the agent reasons and plans. The versioned tool registry defines what the agent can do, and a change to the registry is a change to the agent's capability surface that must be released and recorded as deliberately as a model update. The versioned policy bundle defines the cost caps, step budgets, permissioning, and confirmation gates that bound the agent's behavior. Deploying an agent means deploying a specific combination of these three versioned components, so that any production trajectory can be tied to the exact prompt, tool registry, and policy bundle that governed it, which is what makes an agent reproducible and auditable rather than a moving target.

The named failure modes that agent operators should rehearse are three, and rehearsing them before they occur is the difference between a contained incident and an uncontained one. The runaway loop is an agent that fails to terminate, consuming steps and cost until a cap intervenes, and the rehearsal is verifying that the step budget and hard cost cap actually fire under a synthetic looping condition rather than assuming they will. Tool poisoning is an agent induced, through a malicious input or a compromised data source, to invoke its tools toward an attacker's ends, and the rehearsal is confirming that the permissioning and confirmation gates contain the blast radius even when the agent's reasoning is subverted. Silent partial success is a trajectory that reports completion while having abandoned or partially performed its task, and the rehearsal is auditing trajectories at the boundary of their cost caps to confirm that partial completions are recorded and surfaced rather than reported as successes. An agent operation that has named and rehearsed these three failure modes has done the work that distinguishes a system safe to scale from one accumulating the debt that the earlier chapters of this book warned compounds with every session that runs before the controls are in place.

14.19 Technical Checklist

Apply this checklist as a release gate before declaring any RAG, tool-using, or agentic system production-ready.

1. Each LLM feature is classified as prompt-only, RAG, tool-calling, or agentic, and the matching production playbook is documented and applied.
2. Retrieval quality is measured with documented Recall@K and Mean Reciprocal Rank metrics computed on a maintained evaluation dataset, not only at launch.
3. The empty retrieval rate is monitored continuously, with an alert threshold defined and owned by a named on-call rotation.

4. Embedding model version is stored as a first-class metadata field on every vector in the index.
5. Embedding model upgrades trigger a blue-green re-indexing process with a documented validation procedure and rollback path.
6. The chunking strategy is a versioned artifact with a documented evaluation dataset and the quality metrics used to select it.
7. Hybrid retrieval (dense plus sparse) is evaluated against the production query distribution before selecting dense-only retrieval.
8. Re-ranking is applied to a fixed candidate set with a measured latency budget, not to the full retrieval pool.
9. Every tool integration is either idempotent or has a documented compensating action that reverses duplicate side effects.
10. Tool credentials are scoped to the minimum access required by each tool and reviewed as part of each security assessment.
11. Every agent has a documented step budget, a hard cost ceiling, and explicit termination conditions for each terminal state.
12. Session-level traces capture every step of every agent session with structured events linked by session identifier.
13. Alerts exist for sessions exceeding the step budget, sessions terminated by the cost ceiling, and anomalous tool call patterns.
14. High-consequence, irreversible tool calls require human confirmation during the initial deployment period.
15. Code execution tools run in sandboxed environments with no network or file system access outside a designated scratch directory.
16. Agent long-term memory stores have write policies, expiry policies, and audit logs covering all agent writes.

17. Retrieval events within agent sessions are linked to the session trace by session identifier and step index.
18. Prompt injection risk is addressed by architecturally separating retrieved content from instructional positions within the prompt.

14.20 Team Conversation

Use these questions in team meetings to pressure-test alignment on RAG, tool-using, and agentic system operations.

1. Which of our LLM features are honestly RAG or agentic, and are we applying the matching production playbooks to each, or treating them all as prompt-only systems with extra steps?
2. When did we last measure retrieval quality on production traffic, and do we maintain an evaluation dataset, or are we running the same launch-time dataset that is months out of date?
3. If our embedding model changed today, what would the re-indexing plan look like, what would the estimated cost and downtime be, and do we have a documented rollback path?
4. Are our tool integrations idempotent? If not, do we have compensating actions documented and tested for each non-idempotent integration?
5. Have we ever had an agent run beyond its intended step budget? What did it cost, and what would have happened without an external intervention?
6. If a regulator asked us to reconstruct exactly what an agent did during a specific session from three months ago, could we produce a complete and accurate session trace?
7. Are our tool credentials scoped to minimum access, and when did we last review the permission grants against the actual tool requirements?

8. What is our current prompt-injection posture? Have we tested it against representative injection attempts, and what is the blast radius of a successful injection, given our current tool permissions?
9. Do we have a human-confirmation gate for high-consequence tool calls, and is it implemented as an architectural control or as a model instruction that can be overridden?
10. What is the operational cost of a single agent session at the 95th percentile, and do we have monitoring that would alert before that cost becomes a budget incident?

14.21 Key Takeaway

Most production LLM systems shipping in 2026 are no longer single-prompt systems. They retrieve from corpora, call external tools, and reason across multiple steps. Each of these capabilities introduces a distinct tier of production complexity that does not exist in a single-prompt architecture, and each tier requires a distinct set of engineering controls. Treating a RAG system, a tool-using system, or an agentic system as a single-prompt system with extra features is the architectural decision that guarantees eventual operational failure. The pilot-to-production pathway for each shape requires explicit answers to the production questions that shape introduces: retrieval quality for RAG, idempotency for tool-using systems, and cost ceilings, step budgets, and session traces for agentic systems.

Retrieval quality is a measurable production component, not a property of the model. Embedding model coherence, chunking strategy, hybrid search, and re-ranking each contribute independently to the retrieval quality the system delivers on production traffic, and each degrades independently if not maintained. The operational discipline is to instrument retrieval quality as a continuously monitored production signal from the first day of operation, to maintain the evaluation dataset that makes that measurement meaningful as the corpus and query distribution evolve, and to treat retrieval quality regressions with the same urgency as serving latency regressions. Teams that treat retrieval quality as a launch-time property discover its importance when they debug answer-quality failures months after launch, with no baseline to recover to.

The responsible-by-design principles for tool-using and agentic systems are not optional extensions for high-risk deployments. They are the production engineering baseline for any system that takes actions in the world. Scoped credentials, sandboxed execution, idempotent tool integrations, cost ceilings, step budgets, human-confirmation gates for irreversible actions, and comprehensive session traces collectively define the minimum production posture for a tool-using agent. Teams that build these controls into their architecture from the first deployment are building systems that are safe to scale; teams that defer these controls are accumulating a debt that compounds with every agent session that runs on production traffic before the controls are in place.

15 Case Studies, Industry Patterns, and Reference Architectures

15.1 Opening Scenario

A new VP of engineering joins an organization that has five teams shipping LLM-powered features. The VP inherits no shared vocabulary, no common deployment template, and no agreed-upon measure of what "production-ready" means. Team A is calling managed APIs without rate-limit handling or cost attribution. Team B is self-hosting an open model on a cluster sized for a demo, not for production traffic, and is running at 92% GPU utilization with no autoscaling. Team C has a retrieval-augmented pipeline built over a vector store that has not been re-indexed since the underlying document corpus doubled in size. Team D has deployed an agentic workflow that can call internal APIs without a token budget, an audit log, or a circuit breaker. Team E has a fine-tuned model that has never been evaluated on its target task with a held-out test set.

After three weeks of internal interviews, the VP produces a written assessment. The assessment does not blame any team. Every team made reasonable decisions given the time pressure and ambiguity they faced. The assessment identifies the gaps, ranks them by operational impact, and maps each gap to the book's pattern that closes it. The first gap is the absence of a shared reference architecture: without a template that defines the components, interfaces, and observability requirements, each team built a bespoke system that no other team can inspect, operate, or extend. The second gap is the absence of a maturity model. Without a shared vocabulary for where the organization is and where it is going, each team believes it is doing reasonably well. At the same time, the aggregate picture shows an organization in stage two of five. The assessment becomes the map. The next two quarters of the road map are set in a single afternoon.

This chapter is the toolkit the VP should have started with. The four case studies that follow are composite pictures of the failure shapes and resolution patterns that appear across industries. The reference architecture is the template that would have given each of those five teams a shared starting point. The maturity model is the shared vocabulary for an honest conversation about where the organization actually is. And the ninety-day plan is the concrete action surface that connects the assessment to the outcome.

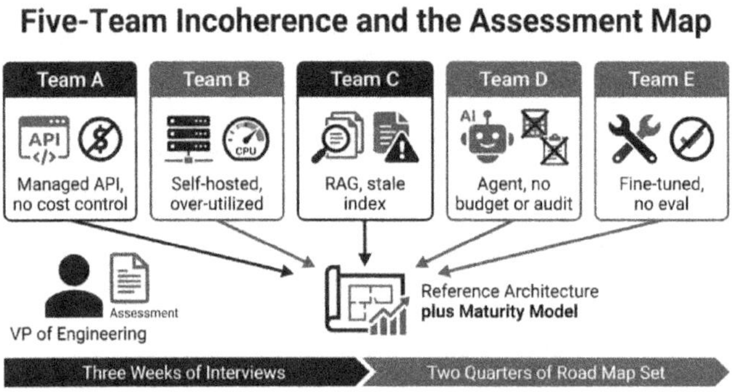

15.2 Why It Matters

Patterns are taught one at a time, but they ship as compositions. The eleven preceding chapters have each addressed a distinct concern: model selection, adaptation, prompt operations, serving, cost, safety, observability, CI/CD, human feedback, and RAG and agentic systems. In any real production deployment, those concerns do not arrive sequentially. They arrive simultaneously, entangled, with the failure mode of one often appearing on the monitoring surface of the other. A latency regression is sometimes a serving issue, sometimes a prompt-length creep, sometimes an agent loop, and sometimes a retrieval-quality collapse that causes the model to process longer, noisier context. Understanding the patterns individually is necessary but not sufficient. Understanding how they compose is what separates a team that can debug production incidents quickly from a team that runs a week-long postmortem to discover that the root cause was a chunking strategy update that was never logged.

The workflow-level impact of composition failures is high. When patterns are not deliberately composed, the gaps between them become the most expensive operational surface in the system. A model without an evaluation suite (model-without-eval) produces no signal when its quality degrades. A prompt managed as application configuration rather than as a versioned artifact (prompt-as-config) has no change history to inspect when behavior shifts. An agent with no token budget (agent-without-budget) has no upper bound on the cost it can accumulate in a single invocation. These are not exotic failure modes. They are the default failure modes of an LLM system built by a team that has not yet read the patterns as a composition.

The case studies in this chapter are not cautionary tales about bad decisions. They are accurate pictures of what mission-aligned production LLM systems look like when the patterns are composed correctly, and what the path from incoherence to operational clarity actually requires. Each case study is drawn from the failure shapes that appear most frequently across financial services, healthcare, e-commerce, and enterprise productivity deployments. The reference architecture at the end of the chapter is the synthesis: the template that makes the composition explicit, the maturity model that gives the team an honest baseline, and the ninety-day plan that connects the baseline to the next stage.

15.3 Case Study: Financial Services LLM Deployment

A mid-sized financial institution had been using a vendor-hosted language model to assist its advisory staff in drafting client communications, summarizing account histories, and generating preliminary research briefs. The system launched under a narrow pilot agreement that exempted it from the firm's full technology review process. When the pilot was declared successful and the system was scaled to the full advisory team, the exemption was not revisited. Within sixty days of full deployment, three incidents forced an emergency review.

The first incident was a hallucinated figure in a client communication. An advisor sent a communication to a client that cited a specific year-to-date return figure for a fund the client did not hold. The figure was plausible. The model had generated it by blending information from two different account summaries in a long context window. The advisor had reviewed the draft but had not verified the specific figure against the source data, because the system had never shown them an incorrect figure in the pilot phase. The pilot corpus was small enough that the context window never held conflicting information. The production corpus was not.

The second incident was a near-miss in compliance. A research brief generated by the system included a recommendation phrasing that, under the firm's regulatory framework, constituted a specific investment advice statement rather than a general educational communication. The distinction is narrow and depends on precise word choice. The model had no instruction that distinguished the two because the product team authored the system prompt without review by the compliance function. The output cleared the advisor's inbox review but was caught by the compliance team's random sampling process, which reviewed one in fifty outbound communications.

The third incident was a data exposure event. A support engineer investigating a latency issue ran a diagnostic query against the serving layer that returned a cached prompt payload containing the personal financial details of eight clients. The caching layer had been added to reduce API costs after the pilot scaling. The security team had not reviewed it, and it stored raw prompt text, including any PII the advisor had included in their query.

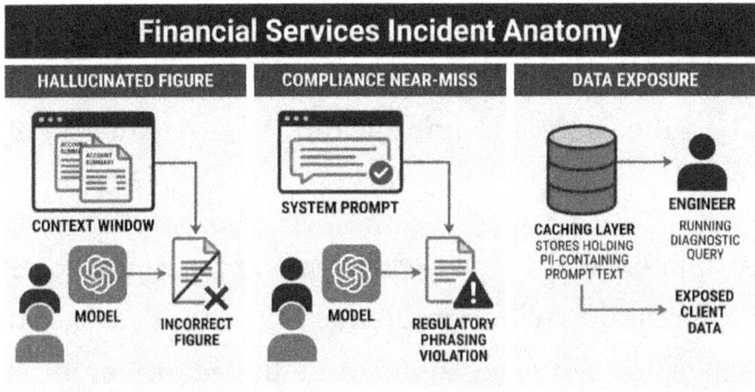

The remediation followed the patterns of the preceding chapters' names. For hallucination control, the team implemented a retrieval-augmented architecture that grounded every draft in specific source documents retrieved from the firm's verified data systems, with citation generation that allowed the advisor to verify each fact against its source. The model was given explicit instructions that it could not state a figure that did not appear verbatim in the retrieved context. An automated faithfulness evaluator, calibrated against human evaluation on a representative sample, was added to the serving path to flag drafts where the model's output diverged from its retrieved sources. Drafts that failed the faithfulness check were held for advisor review with the failing passages highlighted.

For compliance, the system prompt was converted into a versioned artifact owned by a named combination of the product team and the compliance function, with a documented review process that applied to any change. A separate compliance-specific evaluation suite was built covering the regulatory distinctions the firm's framework required, including the investment advice boundary. This evaluation suite ran in the CI/CD pipeline on every prompt change, and any change that caused a new failure in the compliance suite required a compliance review before deployment. The prompt was annotated with the regulatory rationale for each constraint, so that future editors had the context to understand the consequence of a change.

To address the data exposure, the caching layer was removed and replaced with a semantic caching implementation that stored only the embedding vector for each query and the identifier of the cached response, never the raw prompt text. PII detection was added to the prompt assembly path to flag prompts that contain personal financial identifiers. The security review process was updated to include LLM serving layer components as in scope, with specific attention to any caching, logging, or diagnostic access path that touched prompt text. The serving layer's access log was updated to audit any query that accessed a cached payload, and an alert was added for any diagnostic access outside the normal on-call rotation.

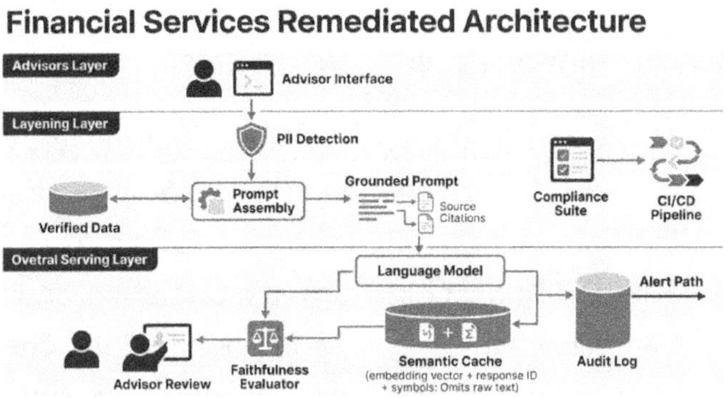

The cross-case lesson from the financial services deployment is that the regulatory and compliance requirements of a domain must be treated as design inputs, not as post-launch approval gates. The pilot exemption that enabled the system to launch without compliance review did not reduce the regulatory requirements; it deferred the cost of meeting them until the system was operating at full scale, where the cost of a compliance failure is highest. The pattern for regulated deployments is to involve the compliance function in the system prompt authoring process from day one, to build the compliance evaluation suite before the first deployment, and to treat any change to the prompt or the retrieval configuration as a compliance event requiring review.

15.4 Case Study: Healthcare NLP Pipeline

A regional health system deployed a language model pipeline to assist clinical documentation specialists in drafting encounter summaries from raw clinical notes. The pipeline was positioned as a drafting assistant: the clinician would review the draft, correct any errors, and approve it before it entered the patient record. The system was not positioned as a diagnostic tool and was not subject to FDA clearance requirements under the existing implementation scope. PHI handling, however, was subject to HIPAA, and the system's data flow introduced several PHI exposure risks that the initial design had not fully mapped.

The primary PHI risk was in the serving layer. The model was hosted by a managed API provider whose business associate agreement the legal team had reviewed for the vendor's core API service, but not for the additional logging and monitoring services the engineering team had enabled to support observability. Those monitoring services retained prompt text, including raw clinical notes, for thirty days by default. The raw clinical notes were not de-identified before being sent to the model. The health system's HIPAA compliance framework required that any PHI transmitted to a business associate be covered by a signed BAA that explicitly named the services to which the PHI was transmitted. The monitoring services were not named.

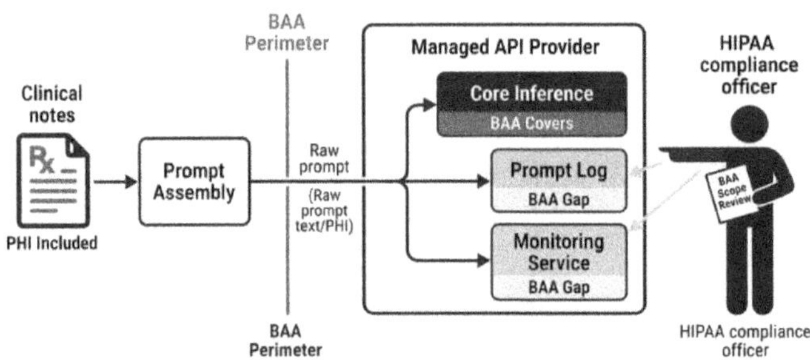

The remediation required both a legal and an engineering response. On the legal side, the team worked with the managed API provider to obtain a BAA amendment that explicitly named the monitoring and logging services, or to turn off those services and implement equivalent monitoring within the health system's own infrastructure. The engineering team chose the second path: they turned off the vendor's prompt logging, implemented a self-hosted observability stack that ingested prompt-level telemetry within the health system's own HIPAA-compliant environment, and applied de-identification to any prompt text before it left that environment for analytical purposes.

On the engineering side, the team added a de-identification step to the prompt assembly path. Clinical notes were passed through a named-entity recognition model trained to identify PHI categories before being included in the prompt sent to the external model. The de-identification step replaced identified PHI with category placeholders (for example, replacing a patient name with a [PATIENT] token and a date of birth with a [DOB] token) before transmission. The de-identified prompt was then sent to the external model. The language model's draft output was post-processed to restore the placeholders with the original values from the clinical note before being presented to the documentation specialist for review.

The clinical evaluation challenge was distinct from the PHI handling challenge. The system produced drafts that were grammatically fluent and structurally consistent with encounter-summary conventions, but clinical accuracy required domain-specific evaluation that the engineering team could not perform. The team established a clinical evaluation panel of five documentation specialists who scored a random sample of one hundred drafts per week on four dimensions: factual accuracy relative to the source note, completeness of clinically relevant information, absence of information not present in the source note, and consistency with the health system's documentation style guide. The evaluation panel's scores were aggregated into a weekly quality signal that the engineering team used to monitor for drift and to prioritize improvements to the system prompt and the retrieval configuration.

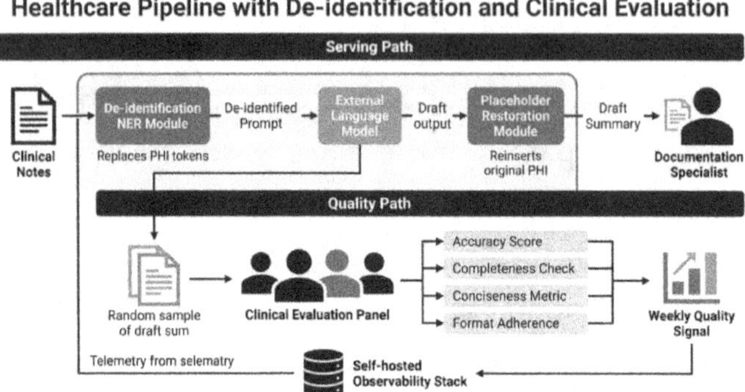

The escalation pattern was the third engineering investment. The system was designed as a drafting assistant, but the team recognized that some clinical scenarios required escalation to a clinical expert rather than a draft. The escalation trigger was implemented as a classifier that ran on the raw clinical note before draft generation: notes that contained certain clinical flags were routed to a human documentation specialist without going through the language model pipeline. The classifier's coverage was defined in consultation with the clinical team and documented as a first-class configuration artifact. The escalation rate was monitored as a production metric, with an alert if the rate deviated significantly from the baseline, because both a rising escalation rate (indicating an increase in complex clinical scenarios) and a falling escalation rate (indicating the classifier might be failing to trigger correctly) were operationally significant signals.

The healthcare case study demonstrates the pilot-to-production pathway for a regulated clinical deployment. The PHI handling requirement, the BAA scope, the de-identification step, the clinical evaluation panel, and the escalation classifier are all elements that must be in place before a clinical documentation system is production-ready. The system that launches without them is not a pilot with known gaps; it is a non-compliant system with unknown exposure. The responsible-by-design pattern for regulated healthcare deployments is to treat PHI handling, clinical evaluation, and human escalation as production requirements that are in scope from the first deployment, not as enhancements planned for a future sprint.

15.5 Case Study: FHIR-Grounded Clinical Decision Support Assistant

Scenario

A regional integrated delivery network operating several hospitals and a network of affiliated clinics deployed a language model assistant to surface relevant patient context to attending clinicians at the point of care. The clinical problem the assistant addressed was well understood: an attending clinician seeing a patient must assemble a coherent picture from a fragmented record that spans encounters, observations, conditions, medications, allergies, and diagnostic reports, and the time spent assembling that picture competes directly with the time available for the patient. The assistant was scoped to answer context-assembly questions, such as summarizing a patient's active conditions and current medications, identifying relevant prior encounters, and flagging documented allergies against a proposed medication. The defining design constraint was that every answer had to be grounded in the patient's actual record rather than generated from the model's parametric knowledge, because an ungrounded clinical assertion is not a helpful summary but a safety hazard. The assistant retrieved from a FHIR-conformant clinical data repository, drawing on Patient, Encounter, Observation, Condition, MedicationRequest, AllergyIntolerance, and DiagnosticReport resources, and it grounded every answer in the specific resources it retrieved. Critically, the assistant surfaced both the answer and the underlying resource references to the clinician, so that the clinician could verify any assertion against the source resource before relying on it. The assistant was positioned as a context-surfacing aid, not a decision-maker, and that positioning shaped every architectural and governance choice that followed.

Architecture

The architecture was organized around a single principle: the FHIR repository was the system of record, and the assistant was a read-only consumer of it. The repository held the canonical patient data, and the assistant never became an alternative store of clinical truth; it retrieved current data for each query rather than maintaining its own copy that could drift from the record. The retrieval layer issued FHIR API calls scoped by SMART-on-FHIR authorization, so that the assistant operated under the same access scopes as the clinician invoking it and could not retrieve data the clinician was not authorized to see. This tied the assistant's data access to the existing clinical authorization model rather than introducing a parallel one, which both simplified the security review and ensured that the assistant inherited the access controls the organization had already validated.

The retrieval design drew a deliberate line between structured and unstructured data. Structured FHIR resources, including conditions, medications, allergies, and observations, were retrieved by code rather than by embedding similarity, because these resources carry coded values whose meaning is exact and whose retrieval should be deterministic. A query about active medications resolved to a structured FHIR query for MedicationRequest resources with an active status, not to a semantic search that might approximate the answer. Embedding-based retrieval was reserved for the unstructured clinical narrative held in DocumentReference resources, where free-text notes carry information that has no coded representation and where semantic retrieval is the appropriate tool. This split meant that the assistant's most safety-critical assertions, about medications and allergies, rested on deterministic structured retrieval, while semantic retrieval was confined to the narrative content where its approximate nature was acceptable.

The language model serving layer sat behind a sandboxed inference gateway that mediated every model call. The gateway enforced that the model received only the retrieved FHIR context assembled for a specific query, applied output controls before returning any response, and isolated the model from any path that could write back to the repository. An audit pipeline ran alongside the serving path and recorded every retrieval and every output, capturing which resources were retrieved for which query and what the assistant returned, in a form designed for HIPAA-compliant logging from the outset. The audit pipeline was not an afterthought bolted onto the serving stack; it was a first-class component through which every interaction flowed, because the governance requirements that follow made comprehensive, immutable logging a precondition for operating the assistant at all.

Governance and Compliance

The governance framework began with HIPAA, because the assistant handled protected health information at every step, and extended into the specific regulatory questions a clinical assistant raises. The 21st Century Cures Act information-blocking provisions shaped the design positively, because an assistant that improves a clinician's access to a patient's own record advances the access goals the provisions promote, but the organization confirmed that the assistant did not inadvertently restrict or obscure information in ways the provisions prohibit. The FDA's evolving stance on clinical decision support software was the central regulatory question, and the assistant's scope was defined to remain on the non-device side of that line: it surfaced and summarized information already in the record with the underlying references exposed, allowing the clinician to independently review the basis for any assertion, rather than providing a specific diagnostic or treatment recommendation that the clinician could not independently verify. Any retrospective evaluation of the assistant's performance that used patient data was routed through institutional review board review, because evaluating a clinical tool against real patient records is research that carries its own oversight obligations.

The organizing governance principle was that the assistant was an aid and never an autonomous actor. The language model never wrote back to the FHIR repository, never placed an order, and never altered the record in any way; its outputs were read-only context delivered to a clinician who retained full decision authority. This assistant-not-autonomous framing was not merely a policy statement but an architectural guarantee, enforced by the inference gateway's isolation of the model from any write path.

The framing resolved the most difficult governance questions before they arose, because a system that cannot act cannot take an unauthorized clinical action, and the accountability for every clinical decision remained unambiguously with the clinician who made it.

What Went Right

The deployment produced a measurable reduction in chart-review time, as clinicians spent less time assembling context manually and more time on the patient encounter, which was the outcome the program was funded to achieve. The grounded-citation design produced an unanticipated benefit in clinician trust: because every assertion carried a reference to the source FHIR resource, clinicians could verify the assistant rather than having to trust it, and that verifiability accelerated adoption among clinicians who would have been justifiably skeptical of an ungrounded summarization tool. The most durable operational win was the model-update protocol the team established, which required that any change to the underlying language model trigger a re-evaluation of FHIR retrieval quality alongside the model evaluation, because the team recognized that answer quality depended on retrieval and model together and that a model change could interact with retrieval in ways a model-only evaluation would miss. A secondary win emerged in incident handling: because every interaction flowed through the audit pipeline, the rare cases in which a clinician disputed an assistant output could be reconstructed exactly, traced to the specific resources retrieved and the context assembled, and resolved as either a retrieval gap or a model error rather than left as an unattributable complaint. That reconstructability shortened the investigation of each disputed case from days to hours and gave the clinical governance committee a factual basis for its periodic review of the assistant's behavior.

Lessons Learned

The first lesson was that FHIR conformance is the data contract that makes a retrieval-grounded clinical assistant tractable at all. Because the repository exposed data through standardized resources with coded values, the retrieval layer could issue precise, deterministic queries against a known schema rather than parsing idiosyncratic data formats, and that contract is what made grounded retrieval reliable enough for clinical use. The second lesson reinforced the architecture's central split: structured FHIR retrieval beat free-text retrieval for clinical accuracy on the assertions that mattered most, and the team came to regard semantic retrieval over narrative notes as a complement to structured retrieval rather than a substitute for it. The third lesson was that audit logging had to be designed in from day one rather than retrofitted, because the governance requirements made comprehensive, immutable logging a precondition for operation, and a logging layer added after the fact would have left a gap in the record for every interaction that preceded it. The fourth lesson was that clinician-in-the-loop was not an optional safeguard but a structural property that changed the operational service-level agreement, because an assistant that surfaces context for a clinician's judgment carries different latency, availability, and accuracy obligations than an autonomous system, and the team designed its SLA around the clinician's workflow rather than around an abstract uptime target.

15.6 Case Study: E-Commerce Recommendation and Search

A large online retailer integrated a language model into its product search and recommendation pipeline to improve the relevance of search results for natural-language queries and to generate personalized explanations for recommendations. The integration replaced the previous keyword-matching search layer with a hybrid retrieval system that combined dense semantic retrieval over product embeddings with sparse BM25 retrieval over product metadata. The language model generated recommendation explanations from a product description template and real-time user context signals. The deployment target was the peak holiday traffic period, which drove approximately eight times the retailer's average daily query volume across a four-week window.

The latency budget was the primary engineering constraint. The retailer's search quality team had measured that each one-hundred-millisecond increase in search result latency reduced conversion by a measurable fraction of a percentage point at their traffic scale. The target end-to-end search latency was two hundred and fifty milliseconds at the ninety-ninth percentile. The hybrid retrieval layer added approximately 80 milliseconds of latency compared to the previous keyword system under normal load. The language model explanation generation added another 110 milliseconds. The total system was at 200 milliseconds at the 90th percentile and 340 milliseconds at the 99th percentile: within budget at the median but over it in the tail.

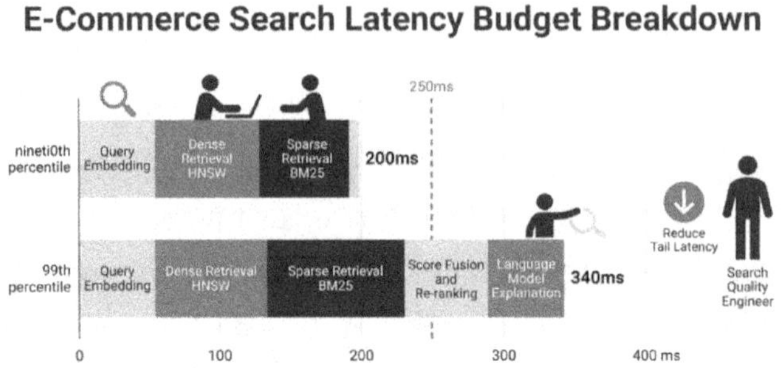

The team used three techniques to bring the ninety-ninth-percentile latency within budget. The first was speculative execution for the language model explanation. For the top-five ranked products, the explanation generation request was sent to the model concurrently with the re-ranking step rather than sequentially after it. Because the re-ranking result was stable for the top five positions in ninety-four percent of queries, the explanation for the most likely top result was already generated by the time re-ranking completed. The speculative result was used when it matched the re-ranking output, and discarded and regenerated when it did not. This reduced the latency of sequential explanations to near-zero for the majority of queries.

The second technique was response caching with a semantic similarity threshold. For product queries in which the embedding distance to a recently cached query was below a defined threshold, the cached explanation was served, with a freshness check against the product catalog. Because popular products generate repetitive queries, the cache hit rate for explanation generation reached 61% at peak traffic, effectively eliminating model latency for the majority of the query volume during the highest-traffic window. The cache TTL was set to 4 hours to balance freshness, cost, and latency.

The third technique was model tiering. The explanation generation path was split into a fast path and a quality path based on the session's estimated value. Sessions with a high purchase probability score (computed from prior session behavior) were routed to a larger, slower, higher-quality model. Sessions with a low purchase probability score were routed to a smaller, faster model whose explanation quality was measurably lower but whose latency was 50 milliseconds rather than 110. The tiering logic was implemented as a simple classification gate at the serving layer, with monitoring of the routing distribution to ensure the quality path was neither over- nor under-utilized.

E-Commerce Hybrid Search and Tiered Explanation Architecture

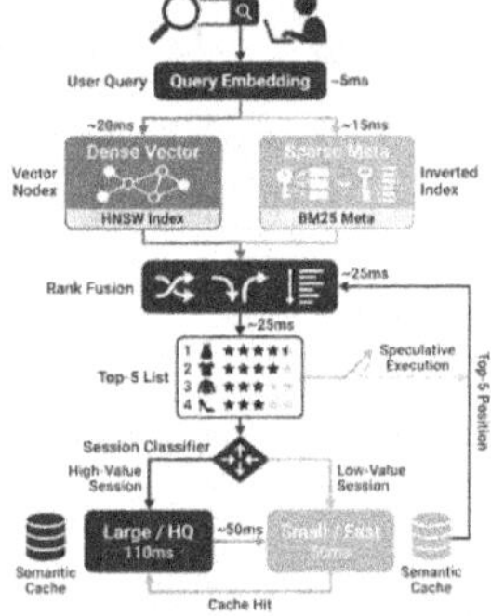

The peak traffic resilience required a separate engineering investment. The retailer's traffic during the peak holiday window was not uniformly eight times average; it was variable, with bursts that could reach twelve to fourteen times average for short intervals when a promotional event fired. The serving cluster was sized for eight times the average, with a twenty-percent headroom buffer, but the burst peaks could saturate the cluster. The team implemented a load-shedding pattern for the explanation generation path. When the serving cluster exceeded a defined utilization threshold, explanation-generation requests were queued with a maximum wait time, and requests that exceeded the wait time were served a fallback response that omitted the explanation and returned only the ranked product list without the language-model content. The fallback degraded the experience but preserved the core search functionality and the latency budget.

The e-commerce case study illustrates three patterns that appear together in high-traffic latency-sensitive LLM deployments: speculative execution to eliminate sequential latency in the serving path, semantic caching to reduce model invocations for repeated or similar queries, and model tiering to match model cost and quality to session value. None of the three patterns is specific to e-commerce; all three apply wherever a latency budget is binding, traffic is variable, and the cost of model invocations is material. The load-shedding pattern for the explanation path is the graceful degradation companion to the latency optimization patterns: it ensures that when the latency budget cannot be met, the system degrades in a controlled way that preserves the user experience for the core function.

15.7 Case Study: Enterprise Productivity Assistant

A software company deployed an internal productivity assistant that integrates with calendar, email, document, and ticketing systems. The assistant was capable of scheduling meetings, drafting email responses, summarizing documents and ticket threads, searching across all connected systems, and executing multi-step tasks such as creating a ticket from a meeting summary and scheduling a follow-up. The assistant used a ReAct-style agentic loop to plan and execute multi-step tasks, with tool-calling interfaces to each connected system. The deployment surface was the entire company: every employee with a company account had access to the assistant from day one of launch.

The first operational incident occurred within the first week. An employee asked the assistant to clean up their email inbox by archiving all emails older than thirty days that had not been replied to. The assistant interpreted this instruction literally and archived 4,700 emails, including emails in active threads where the employee had been CC'd but had not explicitly replied, emails in folders the employee used as reference archives, and several emails from legal and compliance that contained time-sensitive action items. The employee discovered the scope of the action only when they noticed the empty inbox and could not find a specific email. The archiving operation took eleven minutes to execute across approximately one hundred sequential tool calls, none of which had a confirmation step.

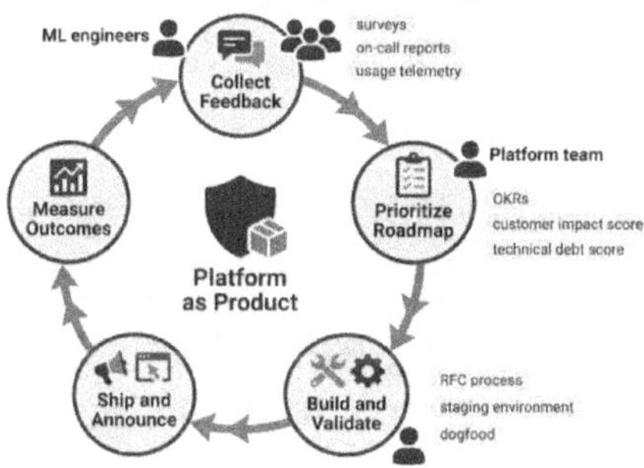

The remediation required changes to the agentic architecture at three levels. The first level was a confirmation gate for high-scope actions. Any tool call that would modify more than a defined threshold of records, send communications to parties outside the employee, or delete or archive content was required to present a confirmation summary to the user before execution. The confirmation summary described the scope of the action in plain language, listed the specific items affected, and required explicit user approval before proceeding. In the archiving scenario, the confirmation summary would have listed 4,700 emails and requested approval before any archiving occurred.

The second level was a tool-call budget per agent invocation. Each invocation of the agentic loop was given a maximum tool-call count, configurable per task type, with a hard ceiling. When the tool-call count for an invocation reached the budget limit, the loop was halted, the state was preserved, and the user was presented with a progress summary and a prompt to approve continuation. The budget prevented runaway loops from executing indefinitely on poorly scoped instructions and created natural human-in-the-loop checkpoints for long-running tasks.

The third level was a scope constraint on the tool-calling interfaces. Each tool-calling interface was updated to enforce a maximum batch size per call, so that a single tool call could not modify more than a defined number of records, regardless of the instruction. The email archive tool was updated to limit the number of emails archived per call, requiring the agentic loop to batch large archiving operations across multiple confirmed steps. This constraint prevented any single tool call from making a large irreversible change before a human had an opportunity to review the scope.

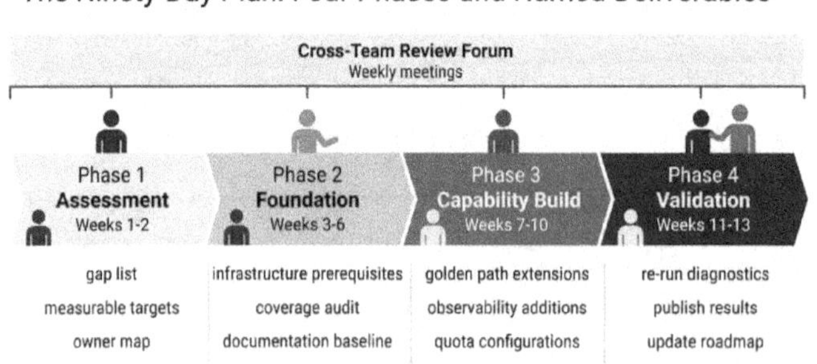

The data-handling requirements for the enterprise assistant presented a second engineering challenge. The assistant had access to email, documents, and tickets containing confidential information, including personnel decisions, customer contracts, unreleased product plans, and legal communications. The prompt assembled for each assistant interaction included content retrieved from these sources, and the language model's training data had no awareness of the content's sensitivity classification. The team implemented a sensitivity classification step in the retrieval path: each document in the connected systems was classified into one of four sensitivity tiers (public, internal, confidential, restricted) based on its metadata and a lightweight classifier. Retrieved content at the confidential and restricted tiers was handled with additional controls: it was not cached, not logged in the standard prompt-output log, and flagged for human review if it appeared in any response shared with a third party.

The enterprise assistant case study illustrates the agentic patterns at the company scale. The confirmation gate, the tool-call budget, and the batch size constraint are the three mechanical controls that make an agentic system safe to deploy to an entire organization without requiring every user to be technically sophisticated about the risks of poorly scoped instructions. The sensitivity classification in the retrieval path is the data handling analog: it ensures that confidential content is handled with additional controls without requiring the user to know which documents in the connected systems are confidential. Both sets of controls implement the responsible-by-design principle: the system's safety properties are enforced by the architecture, not by the user's knowledge of the system's risks.

15.8 Cross-Case Patterns and Lessons Learned

The four case studies share a set of structural patterns that appear across industries and deployment types. Naming these patterns explicitly is useful because it allows a team to recognize them in their own deployment before an incident forces them to do so. The patterns below appear in at least three of the four case studies.

The pilot-exemption trap appears in the financial services and healthcare cases. A system that launches under a pilot exemption does not reduce its regulatory requirements; it defers them until the system is at full scale, where the cost of a compliance failure is highest. The corrective pattern is to treat every LLM deployment, including pilots, as subject to the full compliance, data handling, and evaluation requirements of the target deployment scale, with documented exceptions that are reviewed and renewed rather than tacitly inherited.

The evaluation gap appears in all four cases. No team had an evaluation suite that was both comprehensive and continuously running at the time of the first incident. The financial services team had no faithfulness evaluator. The healthcare team had no clinical quality signal. The e-commerce team had no latency-tail monitoring configured to trigger an alert. The enterprise assistant team had no agent scope monitoring. The corrective pattern is that the evaluation suite is a production artifact that is built in parallel with the system, not after the first incident.

The configuration-as-code gap is evident in financial services and enterprise assistant cases. System prompts managed as application configuration rather than as versioned artifacts have no change history, no review process, and no rollback path. The corrective pattern is to treat every prompt, every tool-calling interface definition, and every retrieval configuration as a versioned artifact with an owner, a review process, and a deployment gate.

The caching-without-security-review pattern appears in the financial services and e-commerce cases. Adding a caching layer to reduce cost or latency without a security review introduces a new surface for data exposure, particularly when the cached content includes user queries or system prompts that contain sensitive information. The corrective pattern is to treat any caching layer as a security-review item with explicit documentation of what is cached, how long it is retained, who can access it, and how it is invalidated.

The compliance-review-as-post-launch-gate pattern appears in three of the four cases. In each case, the compliance function was brought in after the system went live, when the cost of making changes was highest, and the urgency to maintain availability competed with the need to make changes correctly. The corrective pattern is the one this chapter states most explicitly: compliance review is a design input, not a late-stage approval gate. The compliance function must be part of the system prompt authoring process, the evaluation suite design process, and the data flow design process from the first sprint.

15.9 An Anti-Pattern Catalog for Production LLM Systems

The following catalog names the anti-patterns that appear most frequently across production LLM deployments. Each anti-pattern is described with its observable symptom, its root cause, and the corrective pattern from earlier chapters.

Model-without-eval is the pattern where a language model is deployed to production without a continuously running evaluation suite. The observable symptom is that quality degradation is discovered through user complaints or declines in product metrics rather than through a monitoring system. The root cause is treating evaluation as a launch activity rather than a production concern. The corrective pattern is to treat the evaluation suite as a production artifact, built in parallel with the model deployment, with automated evaluation running on every deployment and on a continuous sample of production traffic.

Prompt-as-config is the pattern where system prompts are managed as application configuration rather than as versioned source artifacts. The observable symptom is that prompt changes lack a change history, a review process, and a rollback path, so that when a prompt change causes a quality regression, there is no record of what changed or when. The root cause is treating the prompt as a deployment detail rather than as the primary interface between the model and the product. The corrective pattern is the prompt operations discipline from earlier chapters: versioned prompts, documented rationale for changes, automated evaluation gates, and staged rollout.

Agent-without-budget is the pattern where an agentic loop has no explicit limit on the number of tool calls, the duration of execution, or the scope of actions it can take in a single invocation. The observable symptom is a runaway agent that executes a large number of actions on a poorly scoped instruction, often producing an irreversible state change that the user did not intend. The root cause is treating agentic execution as a deterministic sequential program rather than as a probabilistic planning process that must be bounded. The corrective pattern is to implement a tool-call budget, a confirmation gate for high-scope actions, and a batch size constraint on each tool-calling interface.

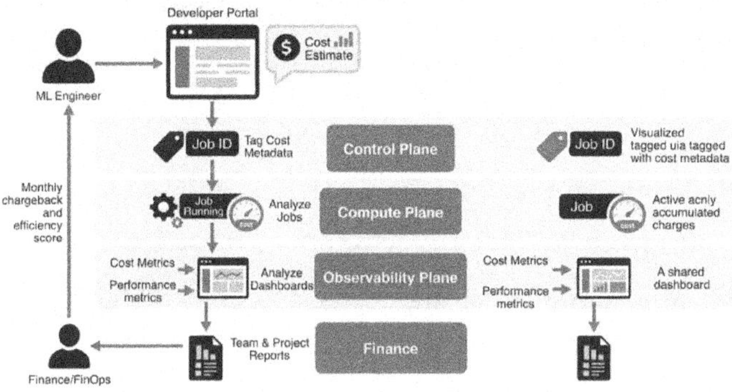

RAG-without-freshness is the pattern where a retrieval-augmented pipeline is deployed without a defined freshness SLA and without monitoring to verify that the source corpus and the vector index are synchronized within that SLA. The observable symptom is that the retrieval returns outdated content that the model presents confidently to the user, because the model has no mechanism to distinguish a stale retrieved document from a current one. The root cause is treating the vector index as a static artifact rather than as a continuously maintained data pipeline with freshness requirements. The corrective pattern is to define a freshness SLA per document category, instrument the ingestion pipeline to monitor and alert on freshness lag, and include the freshness SLA in the system's production-readiness criteria.

Cache-without-review is the pattern in which a caching layer is added to an LLM serving path without a security review of the cache content, retention policy, access controls, or invalidation mechanism. The observable symptom is a data exposure event when a diagnostic query, a monitoring tool, or an unauthorized user accesses the cache. The root cause is treating the cache as an infrastructure optimization rather than as a data store that holds sensitive content. The corrective pattern is to include caching layers in the security review scope, to define explicitly what is cached (ideally only embedding vectors and response identifiers, not raw prompt text), and to apply the same access control and retention policies to the cache as to the source data.

Hallucination-without-grounding is the pattern where a language model is used to generate factual content without a retrieval mechanism that constrains its outputs to verified sources. The observable symptom is model-generated content that is plausible but incorrect, often in a domain where incorrect content has high stakes (financial, clinical, legal). The root cause is deploying a generative model as if it were a retrieval system. The corrective pattern is to use retrieval-augmented generation to ground the model's outputs in specific source documents, to add a faithfulness evaluator to the serving path, and to make the source documents available to the reviewer so that every factual claim can be verified.

15.10 Industry-Specific Compliance Considerations

Compliance requirements shape LLM deployments in industry-specific ways that generic engineering patterns cannot fully anticipate. The following section names the compliance considerations that distinguish each of the four industry domains in this chapter's case studies, along with the engineering patterns that address them.

In financial services, the primary compliance considerations are suitability, auditability, and boundaries of investment advice. Suitability requirements obligate firms to ensure that product recommendations are appropriate for the specific client receiving them; a language model that generates personalized recommendations must demonstrate that its outputs are consistent with the suitability rules that govern the firm's advisory relationships. Auditability requirements obligate firms to produce evidence of the basis for each client communication, which requires that the model's outputs be traceable to specific data sources. Investment advice boundary requirements distinguish regulated advice (which requires specific licensing and documentation) from educational content (which does not); system prompts for financial services deployments must explicitly encode this distinction, and evaluation suites must test for it.

In healthcare, the primary compliance considerations are HIPAA PHI handling, clinical safety, and scope-of-practice boundaries. HIPAA PHI handling requires that any system processing protected health information have a signed business associate agreement with every vendor whose infrastructure touches that PHI, including serving, logging, monitoring, and caching layers. Clinical safety requires that the system's output not substitute for clinical judgment in contexts where that judgment is legally required; the escalation classifier pattern is the engineering implementation of the scope-of-practice boundary. The FDA's Software as a Medical Device framework imposes additional regulatory requirements on systems whose outputs influence clinical decisions; healthcare teams should determine the framework's applicability before deployment, not after.

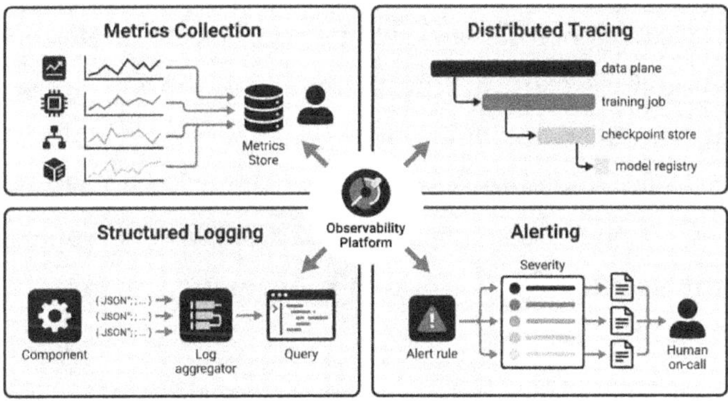

In e-commerce, the primary compliance considerations are consumer protection, transparency in personalization, and disclosure of algorithmic recommendations. Consumer protection requirements in many jurisdictions prohibit dynamic pricing and personalization practices that constitute price discrimination against protected classes; language model personalization systems must be audited to confirm that their outputs do not inadvertently encode protected class proxies from the user's behavioral history. In some jurisdictions, personalization transparency requirements require that users be informed when the content they see is algorithmically personalized. Algorithmic recommendation disclosure requirements for recommender systems are emerging across regulatory frameworks and should be monitored by teams deploying language model recommendation systems.

In enterprise productivity deployments, the primary compliance considerations are data residency, intellectual property and confidentiality, and employment law implications of AI-assisted work. Data residency requirements in some jurisdictions require that employee data, including the queries employees send to an internal assistant, be processed and stored within a specific geographic boundary. Intellectual property and confidentiality requirements govern what content the assistant can include in its context, what it can generate, and what it can export to systems outside the company's control. Employment law implications of AI-assisted work are an emerging area; teams deploying productivity assistants should work with HR and legal to define the boundaries of the assistant's role, particularly in contexts where its outputs influence performance evaluation or resource allocation.

15.11 A Reference-Architecture Template for Production LLM Systems

A reference architecture for a production LLM system is not a diagram. It is a document that lists the components, interfaces, decisions, constraints, and operational commitments the system makes. The diagram is a useful visual companion, but the document is the artifact that can be reviewed, versioned, owned, and used as a production-readiness gate. The template below defines the sections every reference architecture document for a production LLM system should contain.

The first section defines the system's purpose and production. This section states what the system does, who its users are, and what the organization's definition of "production-ready" means for this system. The production definition is not generic; it specifies the latency SLA, the quality threshold, the safety requirements, the compliance framework, and the availability commitment that must be met before the system is declared production-ready, after which any regression is treated as an incident. Teams that skip this section build systems without an agreed-upon definition of success or failure.

The second section is the component map. This section names every component in the system: the model and its serving infrastructure, the retrieval layer (if any), the prompt assembly module, the evaluation suite, the observability layer, the caching layer, the human-in-the-loop interfaces, and the CI/CD pipeline. For each component, the section names the owner, the deployment mechanism, and the interface contract with adjacent components. A component not named in the reference architecture has no owner or interface contract; it is the component that will cause the next incident.

The third section is the decision log. This section records the major architectural decisions and their rationale. The decision log is the document that helps a new team member understand why the system is built the way it is, and it also allows the team to revisit a decision when circumstances change. Each entry in the decision log names the decision, the alternatives considered, the rationale for the chosen option, and the conditions under which the decision should be revisited. Decision logs are particularly valuable for the choices that are not obvious: why a specific model was selected over alternatives, why self-hosting was chosen over a managed API, and why a specific chunking strategy was adopted.

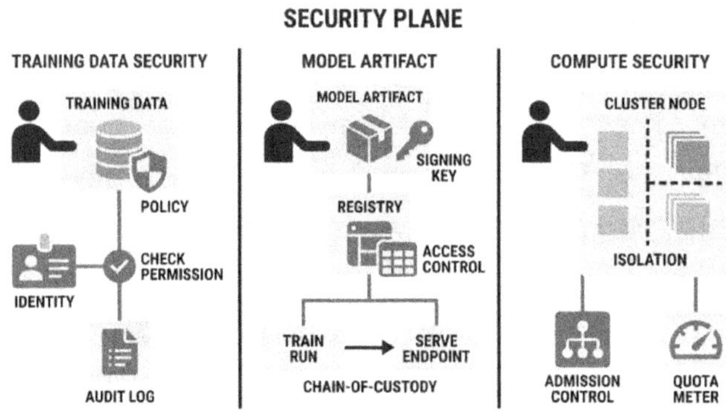

The fourth section is the operational runbook. This section contains the incident playbooks for the failure modes specific to this system, the rollback procedures for each deployable component, and the on-call escalation path. LLM-specific incident playbooks cover hallucination outbreaks, prompt injection findings, drift alarms, and runaway agents. The runbook is a living document that is updated after every incident postmortem. A system without a runbook has documented no operational knowledge from its incidents.

The fifth section is the compliance mapping. This section maps each compliance requirement the system is subject to against the engineering control that addresses it, the owner of that control, and the evidence collection process that demonstrates compliance to an auditor. For regulated deployments, this section is the document that the compliance and legal teams use to satisfy regulatory review. For unregulated deployments, it is the document that ensures the team has thought through the compliance implications of their design.

The sixth section is the metrics commitment. This section names the specific metrics the system commits to in production: the P99 latency target, the quality score threshold, the cost-per-request budget, the safety-rate floor, and any other metric the team treats as a production SLO. For each metric, the section names the measurement method, the alerting threshold, and the owner of the on-call response. A system without explicit metric commitments has no agreed-upon definition of what it means to operate correctly.

15.12 A Maturity Model for Production LLM Capability

The five-stage maturity model below is a tool for honest self-assessment. Its purpose is not to rank teams or to create a leaderboard; it is to give a team a shared vocabulary for where they are and what the next stage requires. The stages are named rather than numbered to emphasize that they describe capability, not progress on an arbitrary timeline. A team at stage two is not behind; they are at stage two, with a clear picture of what stage three requires.

Stage one is Ad Hoc. The organization has deployed one or more LLM features, but deployment decisions are made on a case-by-case basis without shared patterns or standards. There is no shared reference architecture, no shared evaluation framework, and no shared vocabulary for production-readiness. LLM components are treated as black boxes, with their quality assessed informally. Incidents are addressed reactively. Cost is not measured on a per-request or per-feature basis. The primary characteristic of stage one is that each LLM deployment is unique and cannot be inspected, extended, or operated by a team member who did not build it.

Stage two is Repeatable. The organization has developed shared patterns for at least some aspects of LLM deployment, such as a model selection framework, a standard serving configuration, or a shared evaluation approach. Some teams use these patterns; others do not. The organization has a definition of production-readiness, but it is not consistently applied. Incidents are documented, but postmortems do not consistently lead to improvements in shared patterns. Cost is measured at the deployment level but not attributed to specific features or users. The primary characteristic of stage two is inconsistency: some teams are operating well, others are not, and the gap between them is not systematically closed.

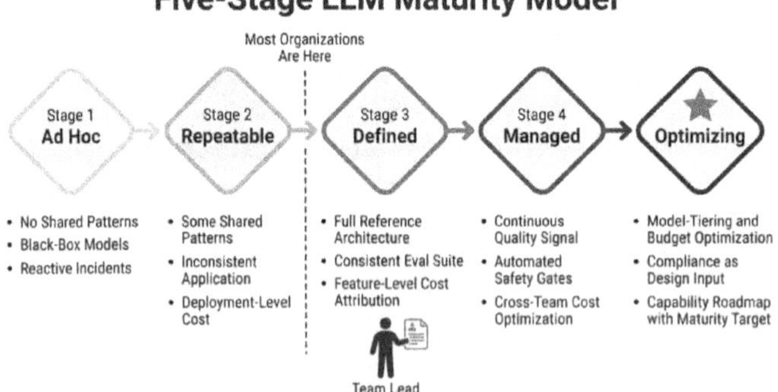

Stage three is defined. The organization has a written reference architecture that is applied to all new LLM deployments. Evaluation suites are built in parallel with every new deployment and run continuously in production. Prompt changes are versioned artifacts with documented change rationale and automated evaluation gates. Cost is attributed to specific features and users. Compliance requirements are identified before deployment and addressed as design inputs. Incidents trigger postmortems that lead to specific improvements in shared patterns. The primary characteristic of stage three is that the organization's LLM practices are documented, applied consistently, and improving based on production experience.

Stage four is Managed. The organization has a continuous quality signal across all production LLM features, with automated alerting on quality degradation. Safety gates are automated and integrated into the CI/CD pipeline. Cost optimization is coordinated across teams rather than optimized independently by each deployment. The compliance function is integrated into the review process for every LLM deployment, including prompt changes and retrieval configuration changes. The organization has a named owner for LLM capability at the platform level. The primary characteristic of stage four is that the organization treats its LLM deployments as a portfolio of production systems with shared infrastructure, shared observability, and coordinated improvement.

Stage five is Optimizing. The organization continuously improves its LLM systems based on production signals, using model tiering, budget optimization, and capability roadmapping explicitly connected to the maturity model. Compliance is a design input with documented patterns that are reused across deployments.

The organization has a public or internal capability roadmap that names the next maturity stage as a target and assigns ownership for the improvements required to reach it. The primary characteristic of stage five is that improvement is the system, not the exception.

15.13 A Ninety-Day Plan to Advance One Maturity Stage

The plan below is written for a team at stage two or three advancing to the next stage. It is structured as four phases of approximately three weeks each, with named deliverables and a single cross-team review forum that provides continuity across phases. The plan is intentionally minimal: it names the fewest deliverables that produce a measurable stage advance, so that a team with limited capacity can execute it without deferring their primary product work.

Phase one is the assessment. The team conducts a structured self-assessment against the target stage criteria, identifying specific gaps between their current practices and the target criteria. The assessment produces a prioritized gap list, with each gap named, its impact on production quality, safety, or cost ranked, and a single owner assigned. The assessment does not attempt to close any gaps; it produces the map. The deliverable is a written assessment document that the team lead presents to engineering leadership at the end of phase one. The forum for this phase is a two-hour assessment session with all team members, facilitated by the team lead, using the maturity model criteria as the agenda.

Phase two is the foundation. The team closes the highest-impact gap identified in the assessment. For most stage-two-to-three transitions, the highest-impact gap is the absence of a written reference architecture. Phase two is spent drafting the reference architecture document using the template from this chapter, with each section assigned to the team member who owns the corresponding component. The draft is reviewed in the cross-team forum at the end of phase two. The deliverable is a reviewed reference architecture document shared with engineering leadership and any compliance or security stakeholders who will use it.

Phase three is the evaluation layer. The team builds and deploys the evaluation suite that will provide the continuous quality signal the target stage requires. The evaluation suite, at a minimum, covers the three highest-risk quality dimensions identified in the phase-one assessment. It runs on every deployment via the CI/CD pipeline and on a continuous sample of production traffic. The cross-team forum reviews the first two weeks of evaluation results at the end of phase three, identifies any evaluation dimensions that need calibration, and confirms that the alert thresholds are producing actionable signals rather than noise. The deliverable is a running evaluation suite with a documented baseline quality score.

Phase four is consolidation. The team updates the maturity assessment in light of the improvements made in phases two and three, confirms the specific criteria of the target stage that have been met, names the criteria that remain open, and produces a prioritized list of the next gaps to close in the subsequent ninety-day cycle. The cross-team forum presents the updated assessment to engineering leadership and secures agreement on the next stage target and the resources required to pursue it. The deliverable is a leadership-approved, updated maturity assessment with a named next stage and a committed owner for the subsequent cycle. The forum is a thirty-minute leadership presentation, not a working session; the content has been prepared in the prior phases.

The ninety-day plan works because it is opinionated about sequence. The assessment before the foundation prevents the team from spending phase two building a reference architecture that addresses the wrong gaps. The foundation before the evaluation layer prevents the team from deploying a continuous quality signal for a system whose component ownership and operational commitments are not yet documented. The evaluation layer before consolidation prevents the team from claiming a stage advance without evidence. And the leadership presentation at phase four creates the organizational commitment that makes the subsequent cycle more likely to be resourced and prioritized.

15.14 Reference Architecture: Multi-Tenant LLM Platform

The multi-tenant LLM platform reference architecture synthesizes the patterns of this book into a single deployable template. It is the architecture appropriate for an organization with multiple product teams consuming LLM capabilities, and that wants to provide those capabilities through a shared platform rather than having each team build and operate its own serving, observability, and cost-attribution infrastructure.

The platform is organized into four layers: the tenant interface layer, the serving layer, the shared services layer, and the infrastructure layer. The tenant interface layer provides each consuming team with a gateway that enforces authentication, applies tenant-specific rate limits, routes requests to the appropriate model or serving configuration, and records the tenant identifier for cost attribution. The gateway is the boundary between the multi-tenant control plane and the per-tenant request context; it ensures that no tenant's request can access another tenant's data or consume another tenant's rate limit.

The serving layer hosts the model serving infrastructure for all tenants. It supports multiple model tiers (a fast, low-cost tier for latency-sensitive use cases and a high-quality tier for complex tasks), continuous batching across tenant requests to maximize GPU utilization, and a semantic cache shared across tenants for queries whose similarity exceeds the sharing threshold and whose response content is not tenant-specific. Model serving configurations are versioned and deployed via the CI/CD pipeline, with a canary deployment pattern that routes a fraction of traffic to the new configuration before full promotion.

The shared services layer provides the capabilities that every tenant needs but that are inefficient for each team to build independently: the prompt evaluation service, the safety filter service, the observability aggregation service, and the cost attribution service. The prompt evaluation service runs automated evaluation on every response in the serving path, applying the faithfulness, safety, and quality evaluators that the platform has validated and calibrated. The safety filter service applies the organization's safety policy to every request and response, with tenant-specific policy overrides for teams whose safety requirements differ from the platform baseline. The observability aggregation service collects telemetry from the serving layer, the evaluation service, and the safety filter and aggregates it into per-tenant dashboards and cross-tenant platform signals. The cost attribution service records the compute consumption of every request by tenant and model tier and produces the chargeback reports that allow the platform to operate on a cost-recovery basis.

The infrastructure layer provides the compute, storage, and networking resources on which the platform operates. GPU allocation is managed by a cluster autoscaler that responds to the aggregate request rate across all tenants, with per-tenant quotas that prevent any single tenant from monopolizing cluster capacity during a traffic burst. Storage is partitioned by tenant for vector indexes and prompt-output logs, with shared storage for the model weights, the evaluation model weights, and the platform configuration artifacts. Network isolation ensures that inter-tenant traffic is not possible within the cluster, and that each tenant's retrieval and prompt assembly operations are scoped to their own data partition.

The multi-tenant platform architecture is the destination for organizations that have advanced past stage three on the maturity model. It requires a platform engineering team, a clear cost-recovery model, a governance process for tenant onboarding, and a shared services layer operated to the same reliability and observability standards as the product services it supports. Organizations that attempt to build a multi-tenant platform before they have a written reference architecture, a continuous evaluation suite, and consistent compliance practices across their teams will find that the platform inherits and amplifies the gaps rather than closing them. The platform is a force multiplier for the patterns; it is not a substitute for them.

15.15 Technical Checklist

Apply this checklist as a release gate before declaring any production LLM system architecture complete against the patterns in this chapter.

1. The team has a written reference architecture document for each production LLM feature, with named owners for each component.
2. Each reference architecture names the patterns from earlier chapters that it composes, with a decision log entry for each major architectural choice.
3. The production definition is written and agreed upon: latency SLA, quality threshold, safety requirements, compliance framework, and availability commitment are all named.
4. Cross-case anti-patterns (model-without-eval, prompt-as-config, agent-without-budget, RAG-without-freshness, cache-without-review, hallucination-without-grounding) have been checked against the team's current architectures.
5. Industry-specific compliance considerations are documented with named owners and evidence collection processes.
6. Any pilot or limited deployment is subject to the same compliance, data handling, and evaluation requirements as the target full-scale deployment.
7. The compliance function is involved in the system prompt authoring process and in the evaluation suite design process, not only in post-launch review.
8. Every caching layer has undergone a security review, documenting what is cached, the retention policy, access controls, and the invalidation mechanism.
9. The team has placed itself honestly on the five-stage maturity model, with specific criteria cited for both the current and target stages.

10. The biggest gap between the current stage and the target stage is the single owner and the target close date.
11. A ninety-day plan includes four phases: named deliverables, a cross-team review forum, and a leadership sign-off event.
12. Engineering leadership has agreed in writing on what the next maturity stage will look like and which resources are committed to reaching it.
13. The operational runbook includes LLM-specific incident playbooks covering hallucination outbreaks, prompt-injection findings, drift alarms, and runaway agents.
14. The metrics commitment section of the reference architecture names specific production SLOs with measurement methods, alerting thresholds, and on-call owners.
15. For multi-tenant platforms: tenant isolation, per-tenant rate limits, cost attribution, and shared safety policy with tenant-specific overrides are all implemented and tested.
16. The evaluation suite has been calibrated against human evaluation on a representative production sample, and its alert thresholds produce actionable signals without chronic noise.

15.16 Team Conversation

Use these questions in team meetings to pressure-test alignment on the patterns in this chapter.

1. Which of the four case studies in this chapter is closest to our most important production LLM feature, and which failures from that case study have we already encountered?
2. Which anti-patterns from the catalog have we shipped at least once, and what did they cost us in engineering time, user trust, or compliance exposure?

3. Where are our industry-specific compliance considerations under-documented, and who is the named owner for closing each gap?
4. What stage of the maturity model are we honestly at, and which specific criteria of the next stage are we farthest from meeting?
5. Where in our stack is incoherence between teams costing us the most, and what shared pattern would close the gap?
6. What is the single most important deliverable in our next ninety days, and who owns it with no ambiguity?
7. If a peer organization asked us how to advance from our current maturity stage to the next, what would the first three weeks of work look like?
8. Does our reference architecture include a written production definition with named SLOs, or is production readiness still an informal judgment?
9. Is our compliance function involved in prompt changes and retrieval configuration changes, or does it see them only after deployment?
10. When we build the next LLM feature, what is the first artifact we will produce, and what will its format, owner, and review process be?

15.17 Key Takeaway

Patterns are deployed in compositions. The eleven preceding chapters have each addressed a distinct concern in isolation. Still, the case studies in this chapter show that production LLM systems encounter those concerns simultaneously and are entangled. The financial services deployment failed at the intersection of prompt governance, faithfulness evaluation, and caching security. The healthcare deployment failed at the intersection of PHI handling, BAA scope, and clinical evaluation. The e-commerce deployment succeeded because the team addressed the latency budget at multiple levels of the serving path simultaneously. The enterprise assistant deployment succeeded after the team addressed agent safety at three architectural levels: confirmation gates, tool-call budgets, and batch size constraints. Understanding patterns in isolation is necessary. Understanding how they compose is what the reference architecture exists to make explicit.

Anti-patterns recur across teams and industries because they share a common root: treating LLM-specific concerns as implementation details rather than as first-class engineering decisions. Model-without-eval treats evaluation as a launch activity. Prompt-as-config treats the prompt as a deployment artifact. Agent-without-budget treats agentic execution as deterministic sequential code. RAG-without-freshness treats the vector index as a static file. Cache-without-review treats the cache as invisible infrastructure. Each of these treatments is understandable under time pressure, but without a shared vocabulary for what "production-ready" means. The reference architecture and the maturity model exist to provide that vocabulary before the incident necessitates it.

Industry compliance is a design input, not a late-stage approval gate. The pilot-exemption trap and the compliance-as-post-launch-gate pattern are the two most expensive compliance anti-patterns this chapter names. Both have the same economic structure: they defer the cost of compliance until the system is at full scale, when the cost of a failure is highest. The responsible-by-design pattern is to bring the compliance function into the design process before the first sprint, to build the compliance evaluation suite before the first deployment, and to treat any change to the prompt, the retrieval configuration, or the data handling path as a compliance event requiring review. A maturity model gives the team an honest baseline and a shared vocabulary for the gap between where they are and where they need to be. A ninety-day plan, structured as four phases with named deliverables and a single cross-team forum, is the concrete action surface that connects the baseline to the next stage. The teams that advance are the ones that treat the maturity model as a production signal and the ninety-day plan as a release gate for the next phase of their LLM capability.

16 Agentic AI Systems in Production

16.1 Opening Scenario

Atlas Logistics, a regional freight and last-mile delivery operator, deployed its first agentic system into production to handle a class of exception-management work that had previously consumed a team of coordinators. The agent was tasked with autonomously resolving delivery exceptions: when a shipment was delayed, refused, or misrouted, the agent would investigate the exception across the company's tracking, routing, and customer systems, decide on a resolution such as rescheduling, rerouting, or notifying the customer, and execute that resolution through the relevant tools. The first week of production began well. The agent handled the routine exceptions, the ones with a clear cause and a single obvious resolution, faster and more consistently than the human coordinators it augmented, and the operations team watched the exception backlog fall for the first time in months.

The first failure mode appeared on the third day. An exception with an ambiguous cause sent the agent into a loop: it investigated, found insufficient information to decide, investigated again along a slightly different path, and continued without converging, accumulating tool calls and model cost against a single exception that a human would have escalated in seconds. The loop did not crash anything; it simply ran, quietly consuming budget, until the per-trajectory cost cap the team had set as a precaution terminated it. The cap worked as designed, but it surfaced the first real operational decision the team had to make: whether to treat the capped trajectory as a failure to be retried or as a signal that the exception was outside the agent's competence and should be routed to a human. The team chose the latter, and in doing so wrote the first line of an operational policy that would grow over the following months. The agent had revealed, in its first week, that operating it was not a matter of model capability but of bounding, observing, and governing a system that acted in the world.

16.2 Why It Matters

Agentic AI is an operations problem distinct from LLM serving, and the distinction is not one of degree but of kind. A served language model produces an output that a caller consumes; the operational surface is latency, cost per call, output quality, and safety of that output, and a bad outcome is a bad answer. An agent takes actions across an unbounded sequence of steps, and its operational surface includes the cost and side effects of every action, the termination of the loop, the recovery of an interrupted trajectory, and the accountability for an action taken without a human in the moment. The serving disciplines established earlier in this book are necessary for agents and insufficient for them, because they bound a single call and an agent is a composition of many calls and many actions whose worst case is not a wrong answer but a sequence of wrong actions executed before anyone notices. An organization that treats its first agent as a more capable served model will discover the gap the first time a trajectory loops, double-acts, or completes the wrong task convincingly.

The business case for agentic systems rests on a single shift: from human-in-the-loop assistance to autonomous workflow completion. An assistant makes a human faster at a task the human still performs and remains accountable for; an agent completes a class of work without a human in the execution path, freeing human attention for the cases the agent cannot handle. The economic appeal is the elimination of routine coordination work, and it is real, but it is conditional on the operational controls that bound the autonomy. The value an agent delivers is proportional to the work it completes autonomously, and the risk it carries is proportional to the same autonomy, which means the engineering discipline of agent operations is not overhead against the business case but the precondition for realizing it safely. A team that cannot bound, observe, and govern an agent cannot responsibly grant it the autonomy that makes it valuable.

16.3 Agent Architectures: From Simple Tool-Using LLMs to Multi-Agent Systems

The simplest agent architecture is a single language model equipped with a set of tools, invoked in a loop until it produces a final answer. The model receives a request, decides whether to call a tool or respond, and if it calls a tool, receives the result and decides again. This architecture is the right starting point for most agentic workloads, because it is the easiest to reason about, observe, and bound. Its trajectory is a single linear sequence of model calls and tool invocations, its cost is the sum of those steps, and its failure modes, while real, are contained within one reasoning thread. A team's first agent should almost always take this shape, because the operational disciplines that the rest of this chapter develops are hard enough to establish on a single-agent system without the additional complexity of coordination.

Multi-step planning agents extend the single-agent pattern by having the model produce an explicit plan before executing it, rather than deciding each step reactively. A plan-execute architecture separates the planning phase, in which the model decomposes the task into a sequence of steps, from the execution phase, in which those steps are carried out and their results observed. The advantage is that an explicit plan is inspectable and can be validated before any action is taken, which matters when actions have side effects, and it makes the agent's intent legible to monitoring rather than emergent from a reactive loop. The trade-off is that a plan made with incomplete information can be wrong, and an agent that executes a stale plan without revising it as observations arrive is brittle, which is why mature plan-execute systems interleave re-planning with execution rather than committing to an initial plan.

Multi-agent systems with explicit role separation decompose a task across several agents, each with a defined role, a scoped tool set, and a bounded responsibility. A common pattern places a coordinating agent that decomposes work and delegates it to specialized agents, each handling a part of the task within its competence, with the coordinator assembling the results. The appeal of role separation is that each agent's capability surface is smaller and easier to permission and reason about, and a specialized agent can be evaluated and bounded independently. The cost is coordination complexity: the trajectory is no longer a single sequence but a graph of interacting agents, the cost and failure surfaces compound across agents, and an error in one agent propagates into the work of others. Multi-agent systems are powerful and are routinely adopted before a team has mastered the single-agent disciplines, which is the most common way an agentic program acquires operational debt it cannot service.

Hierarchical agent systems formalize role separation into layers, with higher-level agents setting goals and delegating to lower-level agents that handle sub-goals, recursively. The hierarchy can mirror the structure of a complex task and can scale to workflows that a single agent could not plan coherently. The architectural trade-off across all of these designs is a single axis: capability against operability. A single tool-using agent is the most operable and the least capable on complex tasks; a hierarchical multi-agent system is the most capable and the least operable, with a failure and cost surface that grows with every agent and every layer. The right architecture is the simplest one that completes the workload, because each step up the capability ladder multiplies the operational surface the team must observe, bound, and govern, and a team that selects an architecture more complex than its workload requires has chosen to operate complexity it does not need.

16.4 The Agent Lifecycle: Design, Evaluation, Deployment, and Retirement

Agent design is a distinct discipline, separate from prompt design and from application engineering, because the artifact being designed is a bounded autonomous behavior rather than a single response or a deterministic program. Designing an agent means specifying the task it completes, the tools it may use, the policies that bound it, the conditions under which it escalates to a human, and the terminal states that end a trajectory. These specifications are design decisions made before any code is written, because they define the agent's competence boundary, and an agent deployed without an explicit competence boundary will be invoked on tasks outside its competence and will fail on them in ways that look like model failures but are design failures. The discipline of agent design is the discipline of deciding, in advance, what the agent is for and what it is not for, and encoding that decision in the agent's tools and policies rather than leaving it to the model's discretion.

Offline evaluation of an agent must be conducted against trajectory test sets, not single-input test cases, because an agent's behavior is a sequence and its failures are sequence-level. A trajectory test set consists of representative tasks paired with the expected behavior across the full trajectory: which tools should be called, in what order, with what arguments, and what terminal state should be reached. Evaluating an agent against such a set surfaces the failures that single-shot evaluation misses, including agents that reach a correct final state through an incorrect or unsafe path, agents that take unnecessary steps, and agents that fail to terminate. The trajectory test set is the agent equivalent of the golden eval set this book establishes for prompts, and it is the artifact that makes agent quality measurable before deployment rather than discovered in production.

Staged deployment for an agent follows the safe-deployment patterns established for serving, adapted to the agent's higher risk. The first stage is shadow deployment, in which the agent runs against real production tasks but its actions are not executed, allowing the team to compare the agent's intended trajectory against the human resolution without any real-world consequence. The second stage is canary, in which the agent executes its actions for a small, bounded fraction of traffic under close observation, so that a failure affects a contained set of cases before the agent is trusted broadly. The third stage is controlled rollout, in which the agent's share of traffic and the scope of its autonomy expand gradually as the team accumulates evidence that it behaves correctly. The staging is more conservative than for a served model because the blast radius of an agent failure is larger, encompassing real actions rather than bad answers, and a team that deploys an agent straight to full traffic has chosen to discover its failure modes at full scale.

Retirement is a lifecycle stage that agent programs frequently neglect, and the retirement signal is as important to monitor as the deployment metrics. An agent should be retired, or its scope reduced, when its trajectory completion rate declines or its human-override rate rises, because both signals indicate that the agent is increasingly being invoked on tasks it cannot complete or whose resolutions humans reject. A declining completion rate means the world has shifted away from the conditions the agent was designed for, and a rising override rate means the humans who supervise the agent no longer trust its outputs, either of which makes continued operation a liability rather than an asset. The discipline is to monitor these signals continuously and to treat a sustained adverse trend as a trigger for re-evaluation and, if warranted, retirement, rather than allowing an agent that has outlived its competence to continue acting because no one defined the condition under which it should stop.

16.5 Tool Registries, Permissioning, and the Capability Surface

An agent's tools are not incidental integrations; they are the agent's capabilities, and they should be managed as versioned, registered, permissioned artifacts rather than as ad hoc API calls scattered through the agent's code. A tool registry is the authoritative catalog of every tool an agent may invoke, recording for each tool its schema, its version, its permissions, and its side-effect profile. Treating tools as registered artifacts means that a change to a tool, including a schema change or a new tool added to the agent's reach, is a deliberate, recorded event subject to review, rather than a silent expansion of what the agent can do. The registry is the artifact that makes the agent's capabilities legible, because the question of what an agent can do in the world is answered by its tool registry and nowhere else.

Least-privilege defaults are the governing principle for tool permissioning, applied to agents with the same rigor that the security chapter applied to credentials. An agent should be granted access only to the tools its task requires, and each tool should be permissioned to the narrowest scope that supports the task, because the agent's potential to cause harm is bounded by the tools it can reach and the scope at which it can use them. The default for any new tool should be no access, with access granted explicitly and recorded, rather than broad access narrowed reactively after an incident. An agent over-permissioned for convenience carries a blast radius far larger than its task requires, and that excess capability is dormant risk that a prompt injection or a planning error can activate at any time.

Capability tokens are the mechanism that enforces permissioning at run time, binding an agent's authority to a specific, scoped, and often time-limited grant rather than to a standing credential. When an agent invokes a tool, it presents a capability token that authorizes that specific invocation within its scope, and the token can encode constraints such as the resources the agent may act on, the operations it may perform, and the window during which the grant is valid. Capability tokens convert permissioning from a static configuration into an enforced run-time control, and they make it possible to grant an agent narrow, just-in-time authority for a specific trajectory rather than broad standing authority that persists between trajectories. The token model also produces an audit benefit, because every authorized action is tied to the specific capability that permitted it.

The capability surface is the security boundary of the agent, and naming it as such is the conceptual contribution of this section. The capability surface is the complete set of actions an agent can take in the world, defined by the union of its tools and their permitted scopes, and it is the boundary within which any agent behavior, whether correct, erroneous, or adversarially induced, is contained. An agent cannot do what its capability surface does not permit, which means the capability surface, not the model's reasoning, is the control that bounds the worst case. This reframes agent safety as a question of capability-surface design rather than of model behavior: a perfectly reasoning agent with an over-broad capability surface is more dangerous than a flawed agent with a tightly bounded one, because the flawed agent's worst case is contained by its boundary while the over-permissioned agent's is not. Designing the capability surface deliberately, keeping it minimal, and reviewing it as the authoritative statement of what the agent can do is the foundational safety discipline of agent operations.

16.6 Multi-Step Evaluation: Trajectory Quality, Not Output Quality

Single-shot evaluation, which scores an agent on the correctness of its final output, misses the failure modes that make agents dangerous, because an agent's risk lives in its trajectory rather than in its final state. An agent can reach a correct final answer through a path that called the wrong tools, took unnecessary actions with real side effects, or succeeded by accident rather than by sound reasoning, and an evaluation that examines only the final output will score all of these as successes. The correctness of the final state is necessary but radically insufficient as a quality measure for a system whose actions, not just its conclusions, have consequences. Agent evaluation must therefore operate at the level of the trajectory, examining the sequence of decisions and actions rather than only the destination.

Trajectory-level evaluation patterns examine the full sequence an agent produces against expectations about that sequence. The evaluation asks not only whether the agent reached the right terminal state but whether it took an appropriate path: whether it called the right tools, avoided unnecessary or unsafe actions, used a reasonable number of steps, and recovered correctly from any errors it encountered. This requires capturing the complete trajectory as the unit of evaluation, including every model decision, tool call, and observation, and scoring that trajectory against a rubric that encodes what good agent behavior looks like for the task. Trajectory evaluation is more expensive and more involved than output evaluation, which is precisely why teams neglect it, and neglecting it is how an agent that passes its output tests fails in production through the path it took rather than the answer it reached.

The "correct final answer for the wrong reasons" failure is the canonical trajectory-level failure and deserves to be named explicitly, because it is invisible to output evaluation and corrosive in production. An agent exhibits this failure when it produces a correct result through reasoning or actions that were not sound, such as guessing correctly, taking an action whose success was coincidental, or reaching a valid state through an invalid path. The failure is dangerous because it passes every output check while being fundamentally unreliable: the same flawed reasoning that happened to produce a correct answer on the test case will produce an incorrect one on a different case, and the agent's apparent competence is an artifact of evaluation that examined only outcomes. Detecting this failure requires examining the trajectory and asking whether the path that produced the answer would generalize, which is a question output evaluation cannot pose.

Rubric-based trajectory scoring is the practical method for evaluating trajectories at scale, and LLM-as-judge is the technique that makes it tractable. A trajectory rubric defines the dimensions of good agent behavior for a task, such as tool-selection appropriateness, action efficiency, error recovery, and adherence to the agent's competence boundary, and each trajectory is scored against those dimensions. Because manually scoring full trajectories does not scale, the same LLM-as-judge pattern this book established for output evaluation is applied to trajectories: a judge model receives the full trajectory and the rubric and produces dimension scores, calibrated against a human-scored reference set to confirm the judge's reliability. Applying LLM-as-judge to trajectories rather than single outputs is the extension that makes trajectory evaluation operationally feasible, and it is the evaluation backbone of a mature agent program, because it lets a team measure trajectory quality continuously rather than auditing a handful of trajectories by hand and hoping they are representative.

16.7 Safety, Human Oversight, and Kill Switches

Every agentic system in production requires a kill switch, and the distinction between hard and soft kill switches determines what a team can do when an agent misbehaves. A hard kill switch halts all agent execution immediately and unconditionally, terminating in-flight trajectories and refusing new ones, and it exists for the case where the agent is causing harm and the priority is to stop it regardless of the partial work abandoned. A soft kill switch is more selective: it prevents the agent from starting new trajectories or from taking new high-consequence actions while allowing in-flight low-risk work to complete or wind down gracefully, and it exists for the case where the team wants to pause the agent without the disruption of an abrupt halt. Both must exist, both must be tested before they are needed, and both must be operable by a human who is not the agent's author, because the situation in which a kill switch is needed is rarely one in which the system is behaving as its designer expected.

Human-in-the-loop checkpoints are the control that keeps a human in the path for high-stakes actions while allowing the agent to proceed autonomously on everything else. A checkpoint pauses a trajectory before a designated action and requires explicit human confirmation before the action executes, converting an autonomous step into a supervised one for the actions where supervision is warranted. The design question is which actions warrant a checkpoint, and the answer is governed by the consequence of the action rather than its frequency: an action that is irreversible, costly, or externally visible warrants a checkpoint even if it is rare, while a routine reversible action does not. A checkpoint implemented as a model instruction the agent is asked to follow is not a control, because the model can fail to honor it; a checkpoint must be an architectural gate that the trajectory cannot pass without the human confirmation, enforced outside the model's discretion.

The autonomy gradient is the framework that organizes oversight by matching the control mechanism to the reversibility of the action, and it is the most useful single concept for designing agent safety. Actions fall into four tiers of increasing consequence. Read-only actions, which retrieve information without changing anything, require no checkpoint and can be fully autonomous. Write-reversible actions, which change state in a way that can be cleanly undone, can be autonomous with logging sufficient to support reversal. Write-irreversible actions, which change state in a way that cannot be cleanly undone, require a human-in-the-loop checkpoint before execution. Destructive actions, which delete or permanently alter critical state, require both a checkpoint and an additional control such as a second approver or a mandatory delay, and many agent designs simply place destructive actions outside the capability surface entirely. The autonomy gradient turns the abstract question of how much to supervise an agent into a concrete classification of each action by reversibility, with a defined control for each tier, and it lets a team grant broad autonomy for low-consequence work while reserving human oversight for the actions where it matters.

16.8 Observability and Tracing for Agent Systems

Distributed tracing, established earlier in this book for multi-step pipelines, is the foundation of agent observability, but an agent trace is more demanding than a pipeline trace because its shape is not known in advance. A served pipeline has a fixed structure that tracing annotates; an agent trajectory has a structure determined at run time by the agent's decisions, branching and looping in ways the trace must capture without a predefined template. The agent trace records every model call, every tool invocation with its arguments and results, every observation, and every plan revision, assembled into a single trace that represents the complete trajectory. This trace is the primary observability artifact of an agentic system, and an agent operated without complete tracing is operated blind, because there is no other way to reconstruct what an agent did across a sequence whose shape was determined by the agent itself.

Per-step latency and cost are the metrics that the trace must carry at the granularity of the individual step, because agent performance problems and cost problems are step-level phenomena. A trajectory that is slow or expensive is slow or expensive because of specific steps, and diagnosing it requires the trace to attribute latency and cost to each model call and tool invocation rather than reporting only the trajectory total. This granularity also enables the cost controls developed later in this chapter, because a system cannot enforce a per-trajectory cost ceiling it cannot measure per step. The instrumentation requirement is that every step in the trace carries its own latency and cost, accumulated into running trajectory totals that the system can act on while the trajectory is still in flight.

The agent trace is a forensic artifact, and designing it as one changes how it is built and retained. When an agent takes an action with a real consequence, whether correct, erroneous, or disputed, the trace is the record that allows the organization to reconstruct exactly what the agent did, why, and under what authority, and that reconstruction is needed for debugging, for accountability, and potentially for regulatory or legal inquiry. A trace built only for performance monitoring will lack the detail a forensic reconstruction requires, such as the exact tool arguments, the capability token under which each action was authorized, and the model decision that led to each step. Designing the trace as a forensic artifact from the outset, with the completeness and the retention that a later inquiry would demand, is the discipline that distinguishes an agent that can be held accountable from one whose actions cannot be fully reconstructed after the fact.

Alerting for agents must operate on trajectory shape, not only on per-call metrics, because the failures that matter are shape-level. A per-call alert fires on a slow or failed model call, which is useful but blind to the agent-specific failures: a trajectory that is taking far more steps than its peers, that is revisiting the same tool repeatedly, that is approaching its cost ceiling, or that has run far longer than a trajectory for its task class should. Alerting on trajectory shape means detecting these patterns in the evolving trace and intervening before the trajectory completes, which is the only point at which intervention can prevent the cost or the action rather than merely report it after the fact. This introduces the long-running-trace problem, which is that an agent trajectory may run for minutes or longer, far beyond the lifetime of a typical served request, and the observability system must handle traces that are open and growing rather than closed and complete, surfacing the state of in-flight trajectories so that an operator can see and act on a trajectory that is going wrong while it is still running.

16.9 Cost, Quotas, and Resource Governance for Agentic Systems

Per-trajectory cost ceilings are the most important single cost control in an agentic system, because the trajectory is the unit at which cost can run away. A trajectory that loops or pursues an unproductive path will accumulate model and tool cost without bound unless a ceiling terminates it, and the ceiling must be enforced against the running cost total that the trace accumulates step by step, terminating the trajectory when the total crosses the threshold. The ceiling converts the open-ended cost of an autonomous loop into a bounded and predictable maximum, which is what makes the agent's cost a planning input rather than an unknowable risk. Setting the ceiling is a design decision informed by the legitimate cost of the task: a ceiling set far above any legitimate trajectory's cost catches only the most extreme runaways, while a ceiling set close to the legitimate maximum catches more failures at the risk of terminating some valid trajectories, and the right setting is the one that bounds the runaway case while leaving headroom for the legitimate worst case.

Per-session and per-tenant quotas extend cost governance to the scopes above the individual trajectory. A session quota bounds the total cost a single conversation or user interaction can accumulate across all its trajectories, which prevents a single session from consuming disproportionate resources even if each individual trajectory stays within its ceiling. A per-tenant quota, in a multi-tenant deployment where the agent serves multiple customers or business units, bounds the cost any single tenant can incur, which both contains the financial exposure from a single tenant's runaway usage and enforces the fairness that multi-tenancy requires. These quotas are the agent-specific application of the chargeback and capacity-planning disciplines this book established, adapted to a workload whose per-unit cost is variable and whose runaway potential is higher than a served model's.

Cost anomaly detection at the trajectory level is the monitoring that catches cost problems the static ceilings and quotas do not, by detecting trajectories whose cost is anomalous relative to their task class rather than merely above an absolute threshold. A trajectory may stay within its ceiling while still costing far more than a typical trajectory for the same task, which is a signal of an inefficiency or an emerging failure mode that a static ceiling would miss. Trajectory-level anomaly detection establishes the normal cost distribution for each task class and alerts when a trajectory deviates from it, surfacing the agent that is solving a problem expensively before the inefficiency scales across many trajectories into a budget incident. This is the cost-observability discipline applied at the granularity the agent workload requires.

The agent-loop-runaway failure mode is the specific failure these controls exist to catch, and naming it clarifies why each control is necessary. A runaway loop is a trajectory that fails to make progress toward termination, repeating investigation or action steps without converging, and it is the agent failure most likely to produce a financial incident because it consumes cost continuously until something stops it. The per-trajectory cost ceiling is the control that stops it, the step budget is the complementary control that bounds it by count rather than cost, the trajectory-level anomaly detection is what surfaces it before the ceiling binds, and the trace is what allows the team to diagnose why the loop occurred. A team that has these controls in place treats a runaway loop as a contained event that fired its ceiling and produced a diagnosable trace; a team without them treats it as a budget incident discovered after the fact, which is the difference the cost-governance discipline of this section is meant to produce.

16.10 Governance, Accountability, and Audit for Autonomous Actions

The accountability gap is the central governance problem of agentic systems: when an agent takes an action, who is accountable for it. A served model produces an output that a human acts on, and the human is accountable for the action; an agent takes the action itself, and the question of accountability has no automatic answer. This gap is not resolved by the observation that the agent acted, because an agent is not an accountable entity, and an organization that cannot answer who is accountable for an agent's actions has deployed a system whose consequences have no owner. Closing the gap is a governance design problem that must be solved before deployment, not a question to be confronted after an agent's action causes harm and the organization discovers that no human was positioned to be accountable for it.

Immutable audit trails for agent actions are the technical foundation of accountability, and immutability is the property that distinguishes an audit trail from a log. Every action an agent takes must be recorded in a trail that captures what was done, when, under what authority, and as part of which trajectory, and that record must be tamper-evident so that it can serve as the authoritative account of the agent's behavior. The audit trail is distinct from the observability trace, though they draw on the same underlying events: the trace serves operations and debugging, while the audit trail serves accountability and must meet a higher standard of integrity and retention because it is the record an inquiry will rely on. An agentic system whose actions are not captured in an immutable audit trail cannot support accountability, because there is no authoritative record of what the agent did to hold anyone accountable against.

Regulatory considerations for autonomous actions are sector-specific and evolving, and the governing principle is that an agent does not change the regulatory obligations that attach to the actions it takes. An action that requires a licensed human's judgment when performed by a person does not become unregulated because an agent performs it; an action subject to record-keeping, disclosure, or approval requirements carries those requirements regardless of whether a human or an agent executes it. The regulatory analysis for an agentic system therefore begins with the regulated nature of the actions in its capability surface, and an agent whose capability surface includes regulated actions inherits the full regulatory burden of those actions, which is frequently a reason to keep the most heavily regulated actions outside the agent's autonomous reach and behind a human checkpoint.

The human-accountable-by-design pattern is the resolution to the accountability gap, and it is an architectural commitment rather than a policy statement. The pattern requires that every consequential action an agent takes is traceable to a specific accountable human, whether through a human-in-the-loop checkpoint that places a person in the decision path for high-consequence actions, or through an accountable owner who is responsible for the agent's behavior within a defined scope of autonomous actions. The design ensures there is never an action for which the answer to "who is accountable" is "the agent," because the agent is not an entity that can bear accountability. Building accountability in by design, by mapping the agent's capability surface to the humans accountable for each tier of action before deployment, is what makes an agentic system governable, and it is the governance discipline that the earlier chapters' treatment of responsible-by-design systems extends into the domain of autonomous action.

16.11 Case Study: Atlas Logistics Autonomous Operations Agent

Six months after the deployment described in this chapter's opening, the Atlas Logistics exception-management agent had settled into a clear division between what scaled and what did not. The routine-exception workload scaled well beyond the team's initial expectations: the agent handled the high-volume, single-cause exceptions, reschedules, address corrections, and standard customer notifications, at a volume that had eliminated the backlog the human coordinators had never been able to clear, and it did so within well-characterized per-trajectory cost ceilings that made its operating cost predictable. The capability surface for these actions was narrow and entirely write-reversible, which meant the agent operated with broad autonomy and minimal human checkpointing on this class of work, and the trajectory completion rate for routine exceptions remained high and stable across the six months.

What did not scale was the ambiguous-exception workload that had produced the agent's first failure in week one. The team had hoped that a more capable model and a richer tool set would extend the agent's competence into the exceptions with unclear causes or multiple plausible resolutions, but the trajectory evaluation told a consistent story: on ambiguous exceptions, the agent's trajectories were longer, more expensive, and more likely to reach the wrong resolution through a plausible-looking path, the "correct final answer for the wrong reasons" failure made concrete. Rather than continue to invest in extending the agent's reach, the team made the ambiguous-exception boundary explicit, routing those cases to human coordinators by design and using the agent's trajectory cost ceiling and step budget as the detector that identified an ambiguous case the agent should not have attempted. The agent became reliably good at the work within its competence boundary precisely because the team stopped trying to push the boundary outward.

One capability was retired outright. An early addition had given the agent a write-irreversible action, the ability to issue a customer credit, behind a human checkpoint, but the checkpoint volume proved high enough that it eliminated the autonomy benefit while the irreversible nature of the action kept the risk, so the team removed the credit-issuance tool from the agent's capability surface entirely and returned that action to the human workflow. The six-month review concluded that the agent's value came from operating a narrow, well-bounded, mostly reversible capability surface at high volume, and that every attempt to expand beyond that surface had cost more in operational complexity and risk than it returned in autonomous completion. The team's roadmap shifted from expanding the agent's competence to deepening its reliability within the boundary it had proven it could operate. The observability investment paid for itself in the same period: the long-running-trace tooling that had felt like overhead during the build surfaced two emerging failure patterns before they became incidents, a tool whose latency had begun to degrade and was about to trip the agent's circuit breaker across many trajectories, and a slow rise in step counts on a subset of routine exceptions that turned out to be an upstream data-quality change rather than an agent regression. Neither would have been visible through per-call metrics alone, and both were caught because the team alerted on trajectory shape and treated the trace as the primary operational artifact rather than as debugging exhaust.

16.12 Technical Checklist

The following checklist is a release gate to verify before declaring an agentic system production-ready against the patterns in this chapter.

1. The agent's task, competence boundary, and explicit terminal states are documented before deployment, and the agent escalates rather than acting outside its competence boundary.

2. The agent architecture chosen is the simplest one that completes the workload, and the choice of single-agent, plan-execute, multi-agent, or hierarchical design is justified against the workload's actual complexity.

3. Every tool the agent can invoke is registered in a versioned tool registry recording its schema, version, permissions, and side-effect profile.

4. Tool permissions follow least-privilege defaults, with no-access as the default and every grant explicit and recorded.

5. Capability tokens enforce permissioning at run time, scoping each authorized action rather than relying on standing credentials.

6. The agent's full capability surface is documented and reviewed as the authoritative statement of what the agent can do in the world.

7. Offline evaluation runs against trajectory test sets, not single-input cases, and scores the full sequence of decisions and actions.

8. Trajectory-level evaluation detects the "correct final answer for the wrong reasons" failure, with rubric-based scoring applied via a calibrated LLM-as-judge.

9. Staged deployment proceeds through shadow, canary, and controlled rollout, with the agent's autonomy expanding only as evidence accumulates.

10. Retirement signals, declining trajectory completion rate and rising human-override rate, are monitored, with a defined threshold that triggers re-evaluation.

11. Both a hard kill switch and a soft kill switch exist, are tested, and are operable by a human who is not the agent's author.

12. Actions are classified on the autonomy gradient (read-only, write-reversible, write-irreversible, destructive), with the matching control enforced for each tier.

13. Human-in-the-loop checkpoints for high-consequence actions are architectural gates the trajectory cannot bypass, not model instructions.

14. Complete distributed tracing captures every model call, tool invocation, observation, and plan revision, with per-step latency and cost.

15. The agent trace is designed as a forensic artifact with the completeness and retention a later inquiry would require.

16. Alerting operates on trajectory shape, including step count, repeated tool use, and cost-ceiling approach, not only on per-call metrics.

17. Per-trajectory cost ceilings are enforced against the running cost total and terminate runaway trajectories.

18. Per-session and per-tenant quotas bound cost at the scopes above the individual trajectory.

19. Trajectory-level cost anomaly detection surfaces trajectories that are anomalous for their task class before they scale into a budget incident.

20. An immutable, tamper-evident audit trail records every agent action with its authority and trajectory context.

21. The human-accountable-by-design pattern maps every consequential action to a specific accountable human, with no action for which the agent is the only accountable party.

22. The named failure modes, runaway loops, tool poisoning, and silent partial success, have been rehearsed against the deployed controls.

16.13 Team Conversation

The following is a condensed working session at Atlas Logistics six months after the agent's deployment, as the platform lead, an

SRE, and the product owner debate expanding the agent's scope. Priya is the platform lead, Arjun is the SRE, and Marcus is the product owner.

Marcus: The routine-exception numbers are strong enough that the operations director is asking why we are not pointing the agent at the ambiguous exceptions too. That is the larger backlog now.

Priya: The trajectory evaluations are the reason. On ambiguous cases the agent reaches a defensible-looking resolution through a path that does not generalize. It looks right and it is right often enough to be dangerous, which is worse than being reliably wrong.

Arjun: I can confirm that from the traces. The ambiguous trajectories are the ones hitting the step budget and the cost ceiling. The controls are catching them, but the fact that they are the ones being caught tells us the agent is out of its competence on that class.

Marcus: So the controls are working. Does that not argue we are safe to expand and let the ceilings catch the failures?

Priya: The ceiling caps the cost. It does not cap the consequence. A capped trajectory on an ambiguous exception still took some actions before it stopped, and on that class we are less confident those actions were right.

Arjun: And every expansion widens the capability surface. The credit-issuance tool is the example. We added it, the checkpoint volume killed the autonomy benefit, and we carried the irreversible risk for nothing until we pulled it.

Marcus: I hear that. What I need to tell the director is not just no, but what we would expand into instead.

Priya: Depth, not breadth. We deepen reliability inside the boundary the agent has proven, rather than pushing the boundary outward. The roadmap is better evaluation and tighter anomaly detection on the work it already does well.

Arjun: I can support that operationally. Expanding the surface means more to trace, more to permission, more to keep accountable. Deepening what exists is work I can actually staff.

Marcus: Then that is the recommendation. We hold the competence boundary, we route ambiguous exceptions to the coordinators by design, and we invest in reliability rather than reach. I can defend that to the director with the trajectory data behind it.

16.14 Key Takeaway

Agent operations is a distinct discipline because an agent is the only LLM shape that acts in the world across an unbounded sequence of steps, and that single property reshapes every operational concern the rest of this book established. Evaluation moves from output to trajectory, because an agent can be right for the wrong reasons. Cost moves from per-call to per-trajectory ceilings, because a loop has no natural bound. Safety moves from output filtering to a capability surface and an autonomy gradient, because the agent's worst case is an action rather than an answer. Observability moves from a closed request trace to a long-running forensic record, and governance moves from a human acting on output to a human accountable by design for an action the agent took. None of these is the serving discipline made harder; each is a discipline the serving chapters never had to address. A team that internalizes this, that bounds, observes, and governs the loop before granting it autonomy, operates agents as a controlled extension of its production practice rather than as a capability it deployed and hoped to contain.

17 Closing

Twelve chapters ago, this book opened with a team staring at the gap between a demo that impressed the executive suite and a production capability that could survive real traffic, real cost pressure, and real regulatory scrutiny. That gap is not a knowledge gap. It is an engineering discipline gap. The gap exists between the first successful prompt and the system that runs that prompt reliably at scale, logs every reasoning trace for audit, degrades gracefully under load, and stays within the budget the finance team approved. Closing that gap is the entire subject of this book.

The journey covered here spans the full lifecycle of a production LLM: selection and adaptation, serving architectures and cost structures, safety posture and observability, human feedback systems, and the frontier patterns of retrieval, agents, and tool use. The platform cycle will rotate. The patterns in this book are a durable investment.

17.1 What You Have Learned

Chapters 1 and 2 established the foundations. Chapter 1 oriented the reader to the production landscape: the taxonomy of model types, how the four deployment models differ on latency, cost, data residency, and governance, what business cases look like when an ROI framework survives a finance review, and how to treat the regulatory environment as a design input. Chapter 2 narrowed the lens to model selection, the most consequential decision an LLM project makes. The capability-versus-cost frontier, latency and throughput profiles across model sizes, and the conditions under which a smaller, faster model outperforms a larger one on a specific task form the vocabulary every later chapter builds on.

Chapters 3 and 4 covered adaptation. Chapter 3 addressed fine-tuning: where it produces durable gains, where it can lead to expensive overfitting, and the PEFT and LoRA reproducibility practices that make an adaptation run auditable and maintainable. Chapter 4 moved to prompt ops: prompt versioning, regression testing, structured-output contracts, and the operator-user separation that prevents a well-designed system prompt from being overridden by an adversarial user input.

Chapters 5 and 6 addressed serving and cost. Chapter 5 covered continuous batching, KV-cache management, speculative decoding, and the multi-tier fallback architectures that prevent a single provider outage from reaching customers. Chapter 6 established that most LLM cost surprises are avoidable. Semantic caching operates as both a cost lever and a quality lever. Token budget contracts, cost tiering by criticality, and the unit-economics framework that translates per-request cost into a business case turn a cost problem into a managed variable.

Chapters 7, 8, and 9 covered the production concerns that the rest of the engineering organization will hold the team accountable for. Chapter 7 addressed the prompt injection surface, hallucination detection and mitigation, data-leakage vectors, and compliance mapping to the EU AI Act and sector-specific frameworks. Chapter 8 built the observability layer with structured reasoning traces, semantic drift detection, and alerting templates. Chapter 9 established evaluation-driven development, golden-dataset curation, and regression harnesses that gate deployments on quality rather than only on uptime.

Chapter 10 addressed the feedback infrastructure that turns user signals into training data and moves preference signals from the production interface back into the adaptation pipeline. LLM systems improve over time because someone designed the feedback path. Chapter 11 covered the frontier patterns: RAG as a technique for grounding model output in a controllable, auditable knowledge source, agentic orchestration, tool-use contracts, and the observability challenges specific to multi-step reasoning chains. Chapter 12 synthesized everything into end-to-end case studies and a five-stage platform maturity model that gives any team a vocabulary for assessing where they are and what to do next.

17.2 The Patterns That Matter Most

Some patterns survive multiple framework cycles. These are worth internalizing deeply, encoding in platform primitives, and defending when a fashionable alternative arrives. The seven patterns below represent the author's judgment about which have proven most resilient across the LLM tooling churn of the past three years.

Evaluation as a first-class artifact is the most important pattern. Every practice in this book depends on the ability to measure whether the system is behaving correctly. An evaluation suite that is not versioned, not gated in CI, and not reviewed when the system changes is a false sense of confidence. Teams that invest in golden datasets and regression harnesses early discover problems before users do.

Tool-neutral architectural seams are the second pattern. The LLM tooling landscape rotates on a two- to three-year cycle. An architecture with clean seams between application logic, the model interaction layer, the retrieval layer, and the observability layer absorbs that rotation with bounded effort. An architecture that mixes concerns across layers requires a rewrite when the tooling changes.

Semantic caching as both a cost lever and a quality lever is the third pattern. A cache returning a validated, high-quality response to semantically equivalent queries reduces token spend and replaces stochastic output with a deterministic one that has already been evaluated. For high-frequency, well-defined queries, the best answer the system has produced is often better than a freshly generated one.

Ground-truth retrieval over freeform generation when stakes are high is the fourth pattern. Retrieving a fact from a controlled knowledge source rather than generating it from model weights is a deliberate choice that trades recall breadth for factual precision, where hallucinations have real consequences. Keeping retrieval and synthesis as distinct roles is what makes the system auditable. Observability with reasoning traces is the fifth pattern. Quality failures are the failure modes that matter most in LLM systems. Structured traces that capture input, retrieved context, chain of thought, and output make those failures visible and debuggable.

Cost tiering by criticality is the sixth pattern. Routing requests to the smallest model capable of meeting the quality requirement for that tier is the architecture that survives budget reviews. The seventh pattern, the maturity model as a planning instrument, converts a sprawling capability backlog into a sequenced program of work with visible progress markers and a clear connection to mission outcomes.

17.3 What Comes Next

Longer-context models are redefining retrieval architecture. Models attending to hundreds of thousands of tokens shift the challenge from recall precision to attention efficiency and cost-per-token management. The cost tiering and semantic caching patterns are the controls that prevent infrastructure costs from scaling with the context window.

On-device and edge inference are moving from specialized applications to mainstream consideration. A model running on a device without network connectivity keeps sensitive data out of the cloud and eliminates per-request API costs. The model selection discipline from Chapter 2 is more important at the edge than in the cloud because the cost of a model too large to fit is measured in shipped hardware. Multi-modal foundation models that jointly process text, images, and audio are arriving in production alongside this shift, requiring evaluation strategies and safety surfaces beyond those needed by text-only pipelines.

Agent ecosystems are maturing into production infrastructure. Multi-agent coordination, tool-use contracts, and state management architectures will grow in importance as organizations deploy agents that collaborate across longer time horizons. The failure surface for multi-agent systems is the product of individual agent failure surfaces, and interaction effects can produce emergent behaviors invisible in any single trace. Regulatory maturation is accelerating alongside these technical shifts. Teams that treat compliance as a design layer rather than a feature will adapt more readily as enforcement intensifies.

17.4 A Final Word to Architects

Operational clarity is the architect's primary obligation: the state in which every person on the team knows what the system is supposed to do, how to tell whether it is doing it, and what to do when it is not. Operational clarity is not achieved by writing a document. It is achieved by building a system that makes its own state legible through structured observability, fails loudly and specifically when something is wrong, and recovers gracefully because recovery was designed in rather than retrofitted.

Mission-aligned design means technical choices connect back to the reason the system was built. When a new technique is more impressive on a benchmark but does not improve any mission-aligned metric, the right answer is to pass on it. The courage this work requires is the courage to ask that question clearly and let the answer be no. The most consistent failure mode across LLM production programs is not under-investment in technology but over-investment in complexity at the wrong stage of maturity. Advance one stage at a time, with named deliverables and a single owner for each gap, and the platform compounds. Skip stages, and the platform accumulates debt faster than it can handle.

The engineers, architects, and technical leads who build reliable LLM systems are not supporting the organization's mission. They are the mission. The model that summarizes a patient's medical history, extracts material terms from a contract, or assists a developer in writing safer code does none of those things reliably without the production engineering discipline behind it. Build systems that earn trust by being observable, testable, and recoverable. Build platforms that outlast the engineers who built them. That is the standard this field deserves.

17.5 Where to Go From Here

The most immediate next step is not to read another book. It is to audit the system you own against one chapter of this book and produce a specific, actionable gap list. Pick the chapter that names the problem your team is currently experiencing. Work through the technical checklist at the end of that chapter against your actual system. The gap between the checklist and reality is the most valuable backlog item you can bring to your next planning session.

The companion repository contains the practical artifacts referenced throughout this book: serving configuration templates annotated with the decisions they embody, prompt library templates with evaluation suites, PEFT training scripts, pre-configured monitoring dashboards, RAG and agentic workflow reference implementations, and the maturity self-assessment worksheet from Chapter 12. These artifacts are tool-neutral and structured to be adapted rather than copied verbatim. The README in each subdirectory explains which patterns the artifact illustrates and what modifications to expect for a specific environment.

For ongoing learning, the MLSys, OSDI, and SOSP conference proceedings publish the research that becomes the next cycle's production patterns. The MLOps Community and LLM practitioners tracks at major ML conferences produce practitioner-grade knowledge that moves faster than any publication cycle. Appendix D contains a curated reading list organized by chapter and topic, intentionally kept short: each entry is included because it has shaped how practitioners think about the problem it addresses.

The highest-leverage internal investment for technical leads is the design review process. A review that asks the same structured questions for every new LLM workload, explicitly names trade-offs, and treats evaluation coverage and observability as non-optional criteria will raise the team's average output quality faster than any individual training program. The discussion questions at the end of each chapter are a starting point for that review template. Models will be superseded. Frameworks will be replaced. The evaluation discipline and observability practice built into the review process will compound across every subsequent platform generation.

www.ingramcontent.com/pod-product-compliance
Lightning Source LLC
LaVergne TN
LVHW010555100826
845148LV00014B/2717

* 9 7 9 8 9 0 4 9 8 0 1 5 3 *